THE ILLUSTRATED FLORA OF ILLINOIS

The Illustrated Flora of Illinois

ROBERT H. MOHLENBROCK, General Editor

THE ILLUSTRATED FLORA OF ILLINOIS

FLOWERING PLANTS
lilies to orchids

Robert H. Mohlenbrock

SOUTHERN ILLINOIS UNIVERSITY PRESS
Carbondale and Edwardsville

FEFFER & SIMONS, INC.
London and Amsterdam

Printed in the United States of America
Designed by Andor Braun
Standard Book Number 8093–0408–2
Library of Congress Catalog Card Number 69–16118

This book is dedicated to
the Author's parents
who provided much support and encouragement
during his developing years
as a botanist

CONTENTS

ILLUSTRATIONS

FOREWORD

After having worked with various aspects of the Illinois Flora for over a decade, I came to the realization that not a great amount of information was known about all the Illinois Flora, and that which was known was of a nature which was virtually useless to the average person wanting to know about the plants of this state. Thus the idea was conceived to attempt something that had never before been accomplished for any one of the United States—a multivolumed flora of the state of Illinois, to cover every group of plants, from algae and fungi through flowering plants. In addition to an account with keys of every plant known to occur in Illinois, there would be provided illustrations showing the diagnostic characters of each species.

An advisory board was set up in 1964 to screen, criticize, and make suggestions for each volume of The Illustrated Flora of Illinois during its preparation. The board is composed of taxonomists eminent in their area of specialty—Dr. Gerald W. Prescott, Michigan State University (algae), Dr. Constantine J. Alexopoulos, University of Texas (fungi), Dr. Aaron J. Sharp, University of Tennessee (bryophytes), Dr. Rolla M. Tryon, Jr., The Gray Herbarium (ferns), and Dr. Robert F. Thorne, Rancho Santa Ana Botanical Garden (flowering plants).

This author is editor of the series, and will prepare many of the volumes. Specialists in various groups will be asked to prepare the sections of their special interest.

There is no definite sequence for publication of The Illustrated Flora of Illinois. Rather, volumes will appear as they are completed.

Robert H. Mohlenbrock

February 18, 1969
Carbondale, Illinois

THE ILLUSTRATED FLORA OF ILLINOIS

FLOWERING PLANTS
lilies to orchids

County Map of Illinois

Introduction

This is the second of many volumes devoted to the flowering plants of Illinois. The first six of these will be concerned with that group of flowering plants referred to by botanists as monocotyledons, or monocots. Technically, these are plants which produce, on the germination of the seed, a single "seed-leaf," or cotyledon, as opposed to dicotyledons, or dicots, which produce two "seed-leaves" upon germination of the seed. More practically, monocots are usually recognized by their long, slender, grass-like leaves and their flower parts often in threes or sixes, although this is not always the case.

Monocots include many plants familiar to most of us, such as lilies, irises, orchids, rushes, grasses, and sedges. The first volume begins with flowering rushes, follows with arrowheads, pondweeds, duckweeds, and cat-tails, and ends with spiderworts and rushes. The second volume commences with lilies, follows with irises, and concludes with orchids. The third and fourth volumes will be devoted to grasses, while the fifth and sixth volumes will include the sedges.

The nomenclature followed in this volume has been arrived at after lengthy consultation of recent flores and monographs of the plants concerned. The sequence of taxa presented here essentially follows that proposed for the forthcoming Flora North America series. It is basically a compromise of the systems of Cronquist (1968) and Thorne (1968). Synonyms, with complete author citation, which have applied to species in the northeastern United States, are given under each species. A description, while not necessarily intended to be complete, covers the more important features of the species.

The common name (or names) is that which is used locally in Illinois. The habitat designation is not always the habitat throughout the range of the species, but only for it in Illinois. The overall range for each species is given from its northeastern to northwestern extremities, south to its southwestern limit, and eastward to its southeastern limit.

The range has been compiled from various sources, including examination of herbarium material. A general statement is given concerning the range of each species in Illinois. Dot maps showing county distribution of each monocot in Illinois are provided. Each dot represents a voucher specimen deposited in some herbarium. There has been no attempt to locate the dot with reference to the actual locality within each county.

The distribution has been compiled from field study as well as herbarium study. Herbaria from which specimens have been studied are the Field Museum of Natural History, Eastern Illinois University, the Gray Herbarium of Harvard University, Illinois Natural History Survey, Illinois State Museum, Missouri Botanical Garden, New York Botanical Garden, Southern Illinois University, the United States National Herbarium, the University of Illinois, and Western Illinois University. In addition, a few private collections have been examined.

Each species is illustrated, showing the habitat as well as some of the distinguishing features in detail. Miriam Wysong Meyer has prepared all of the illustrations.

Several persons have given invaluable assistance in this study. Dr. Robert F. Thorne of the Rancho Santa Ana Botanical Garden and Mr. Floyd Swink of the Morton Arboretum have read and commented on the entire manuscript. Dr. John D. Freeman, Auburn University, has made valuable comments on the genus *Trillium*. For courtesies extended in their respective herbaria, the author is indebted to Dr. Robert A. Evers, Illinois Natural History Survey, Dr. G. Neville Jones, University of Illinois, Dr. Glen S. Winterringer, Illinois State Museum, Dr. Arthur Cronquist, New York Botanical Garden, Dr. Jason Swallen, the United States National Herbarium, Dr. Lorin I. Nevling, the Gray Herbarium, Dr. Robert Henry, Western Illinois University, Dr. John Ebinger, Eastern Illinois University, and Drs. George B. Van Schaack and Hugh Cutler, the Missouri Botanical Garden.

Southern Illinois University provided time and space for the preparation of this work. The Graduate School and the Mississippi Valley Investigations and its director, the late Dr. Charles Colby, all of Southern Illinois University, furnished funds for the field work and the salary for the illustrator.

A detailed discussion of the morphology of Illinois monocots and the habitats of Illinois monocots may be found in the volume covering flowering rushes to rushes.

SEQUENCE OF MONOCOT FAMILIES

The sequence of families of monocots and their placement into orders in most floras follow the arrangement of Engler or modifications of it. Thus the traditional division of monocots in Illinois is into nine orders. Detailed study of monocots during the past few decades by botanists has given us a new perspective on the phylogenetic relationship of the monocots.

The sequence of monocots presented in this series essentially follows that which will be employed in the forthcoming Flora North America. It reflects the thinking of two of the leading phylogenists in the country, Arthur Cronquist and Robert F. Thorne, who have devoted many years to the study of family relationships.

So that comparison may be made between the traditional arrangement of monocots and the arrangement found in this work, the two arrangements are outlined below.

Traditional Arrangement of Illinois Monocots	*Arrangement of Illinois Monocots in this Work*
Pandanales	Alismales
Typhaceae	Butomaceae
Sparganiaceae	Alismaceae
Alismales	Hydrocharitaceae
Najadaceae	Zosterales
Potamogetonaceae	Scheuchzeriaceae
Juncaginaceae	Potamogetonaceae
Alismaceae	Ruppiaceae
Butomaceae	Zannichelliaceae
Hydrocharitales	Najadales
Hydrocharitaceae	Najadaceae
Graminales	Arales
Gramineae	Araceae
Cyperaceae	Lemnaceae
Arales	Typhales
Araceae	Sparganiaceae
Lemnaceae	Typhaceae
Xyridales	Commelinales
Xyridaceae	Xyridaceae
Commelinaceae	Commelinaceae
Pontederiaceae	Pontederiaceae
Liliales	Juncaceae
Juncaceae	Cyperaceae
Liliaceae	Poaceae
Iridales	Liliales
Dioscoreaceae	Liliaceae
Amaryllidaceae	Smilacaceae
Iridaceae	Dioscoreaceae

Traditional Arrangement of Illinois Monocots (continued)	*Arrangement of Illinois Monocots in this Work (continued)*
Orchidales	Iridaceae
Burmanniaceae	Burmanniaceae
Orchidaceae	Orchidales
	Orchidaceae

The Alismales is considered first in this work under the belief that the most primitive families possessed more than one free carpel and a petaloid perianth. The primitive condition of the ovary and fruit is responsible for placing the Butomaceae before the Alismaceae. The Hydrocharitaceae, with inferior ovaries, is considered to be advanced in the order.

The traditional view that the Typhaceae, with its sparse perianth and reduced stamens and ovules, is most primitive among the monocots is not substantiated by morphological evidence. To the contrary, it is thought that the reduced number of flower parts usually indicates an advanced condition.

The next orders, Zosterales and Najadales, are the climax of the plants with free carpels which began with the Alismales. It would seem logical for the Juncaginaceae to follow the Alismales, the major difference being the uniformity of the two series of the perianth in the Juncaginaceae. The climax condition is seen in the reduction of all flower parts and in the aquatic habitat. The recognition of the genera *Potamogeton*, *Zannichellia*, and *Najas* into separate families is backed up by the morphological studies of Miki (1937) and Uhl (1947).

The Arales is an order composed of two families in Illinois. The Sparganiaceae and Typhaceae, following in the Typhales, seemingly are merely wind-pollinated aroids.

According to the sequence followed in this work, the Commelinales, which follow, are composed of the Xyridaceae, Commelinaceae, Pontederiaceae, Juncaceae, Cyperaceae, and Poaceae. These last two families are so large that they are treated in separate volumes in this series.

The Liliales, as constituted here, is composed of Engler's Liliales and Iridales. The order represents the beginning of the development of a perianth whose members are mostly uniform (frequently petal-like) and generally fused together, at least at the base.

It is within the Liliaceae that the classification followed in this work differs markedly from the traditional viewpoint. Not only are the usual liliaceous genera included (except *Smilax*), but so too are the traditional genera assigned to the Amaryllidaceae. *Smilax* is removed from the Liliaceae and placed in the Smilacaceae on the basis of the vining habit and the unisexual (in Illinois) flowers.

The remaining families assigned to the Liliales—Dioscoreaceae, Iridaceae, Burmanniaceae—all possess inferior ovaries. A tendency toward zygomorphy is climaxed in the Orchidaceae, the sole representative of the Orchidales.

HOW TO IDENTIFY A MONOCOT

A key to the identification of the monocot families in Illinois begins the systematic section of this volume. By use of this key, the families may be determined. The reader should then proceed to the family, where a key to the genera of that family in Illinois is provided.

Once the genus is ascertained, the reader should turn to that genus and use the key provided to the species of that genus if more than one species occurs in Illinois. Of course, if the genus is recognized at sight, then the general keys should be by-passed.

The keys in this work are dichotomous—*i.e.*, with pairs of contrasting statements. Always begin by reading both members of the first pair of characters. By choosing that statement which best fits the specimen to be identified, the reader will be guided to the next proper pair of statements. Eventually, a name will be derived.

Key to the Families of Monocotyledons in Illinois

* THOSE ENTRIES MARKED WITH AN ASTERISK APPEARED IN THE VOLUME OF FLOWERING PLANTS COVERING FLOWERING RUSH TO RUSHES.

** THOSE ENTRIES MARKED WITH A DOUBLE ASTERISK WILL BE FOUND IN SUBSEQUENT VOLUMES.

---

1. Plants climbing or twining (if erect, then usually with a few weak tendrils from the upper axils); leaves net-veined; flowers unisexual.
   2. Inflorescence umbellate; ovary superior; fruit a berry____ **Smilacaceae,** p. 128
   2. Inflorescence glomerulate or paniculate; ovary inferior; fruit a capsule_____________________________ **Dioscoreaceae,** p. 146
1. Plants erect or floating in water (tendrils never present); leaves mostly parallel-veined; flowers bisexual or unisexual.
   3. Plants with one or two whorls of leaves.
      4. Flowers radially symmetrical; ovary superior; stamens 6.
         5. Plants never over 50 cm tall; flowers usually borne singly _____________ *Trillium* and *Medeola,* in **Liliaceae,** p. 99
         5. Plants more than 50 cm tall; flowers usually more than one______________________ *Lilium,* in **Liliaceae,** p. 27
      4. Flowers bilaterally symmetrical; ovary inferior; stamen 1____ _________________________ *Isotria,* in **Orchidaceae,** p. 244
   3. Plants with leaves alternate, opposite, basal, or none.
      6. Flowers crowded together on a spadix, often subtended by a spathe _____________________________________ **Araceae** *
      6. Flowers not crowded on a spadix (in *Ruppia,* two flowers are borne on a spadix-like structure.)
         7. Plants thalloid, floating in water__________ **Lemnaceae** *
         7. Plants with roots, stems, and leaves, aquatic or terrestrial.
            8. Perianth absent, or reduced to very minute scales (lodicules) or bristles.
               9. Each flower subtended by one or more sterile scales; plants generally not true aquatics.

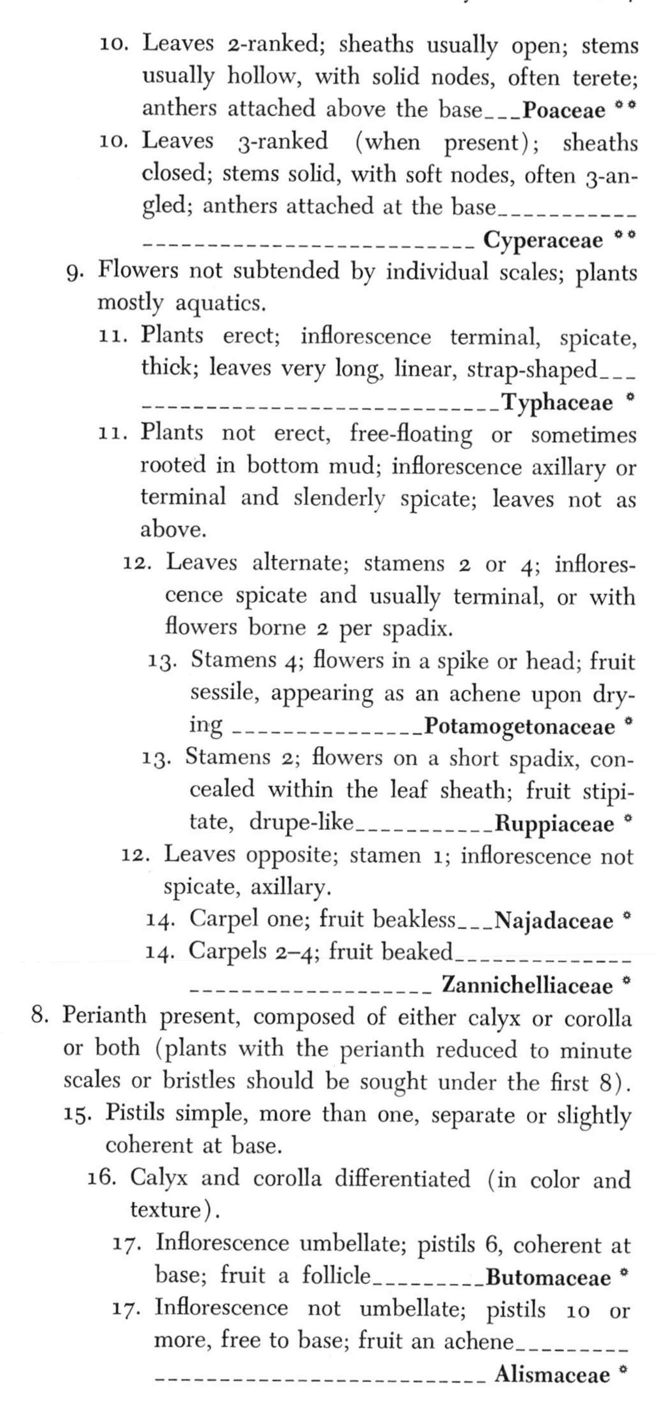

10. Leaves 2-ranked; sheaths usually open; stems usually hollow, with solid nodes, often terete; anthers attached above the base___**Poaceae** **

10. Leaves 3-ranked (when present); sheaths closed; stems solid, with soft nodes, often 3-angled; anthers attached at the base___________ ___________________________ **Cyperaceae** **

9. Flowers not subtended by individual scales; plants mostly aquatics.

11. Plants erect; inflorescence terminal, spicate, thick; leaves very long, linear, strap-shaped___ _______________________________**Typhaceae** *

11. Plants not erect, free-floating or sometimes rooted in bottom mud; inflorescence axillary or terminal and slenderly spicate; leaves not as above.

12. Leaves alternate; stamens 2 or 4; inflorescence spicate and usually terminal, or with flowers borne 2 per spadix.

13. Stamens 4; flowers in a spike or head; fruit sessile, appearing as an achene upon drying _______________**Potamogetonaceae** *

13. Stamens 2; flowers on a short spadix, concealed within the leaf sheath; fruit stipitate, drupe-like___________**Ruppiaceae** *

12. Leaves opposite; stamen 1; inflorescence not spicate, axillary.

14. Carpel one; fruit beakless___**Najadaceae** *

14. Carpels 2–4; fruit beaked______________ ___________________ **Zannichelliaceae** *

8. Perianth present, composed of either calyx or corolla or both (plants with the perianth reduced to minute scales or bristles should be sought under the first 8).

15. Pistils simple, more than one, separate or slightly coherent at base.

16. Calyx and corolla differentiated (in color and texture).

17. Inflorescence umbellate; pistils 6, coherent at base; fruit a follicle_________**Butomaceae** *

17. Inflorescence not umbellate; pistils 10 or more, free to base; fruit an achene_________ ___________________________ **Alismaceae** *

16. Calyx and corolla undifferentiated (*i.e.*, similar in color and texture) ________ **Juncaginaceae** *
15. Pistil one, compound.
18. Ovary superior.
19. Calyx and corolla differentiated (in color and texture).
20. Flowers crowded together in a dense head; leaves basal ______________ **Xyridaceae** *
20. Flowers borne in cymes or umbels; leaves cauline ______________ **Commelinaceae** *
19. Calyx and corolla undifferentiated (*i.e.*, similar in color and texture).
21. Flowers unisexual.
22. Leaves net-veined; flowers in umbels ______________ **Smilacaceae,** p. 128
22. Leaves parallel-veined; flowers in globose clusters, racemes, or panicles.
23. Perianth small, greenish; flowers aggregated in dense globose clusters; stamens 5 ___________ **Sparganiaceae** *
23. Perianth usually conspicuous, greenish, yellowish, white, or bronze-purple; flowers in racemes or panicles; stamens 6 _______________ **Liliaceae,** p. 11
21. Flowers bisexual.
24. Perianth scarious ________ **Juncaceae** *
24. Perianth petaloid.
25. Stamens 3 ________ **Pontederiaceae** *
25. Stamens 6 (or 4).
26. Stamens of different sizes ______________ **Pontederiaceae** *
26. Stamens all alike.
27. Leaves evergreen, rigid; stems woody ___ *Yucca,* in **Liliaceae,** p. 115
27. Leaves deciduous, mostly not rigid; stems herbaceous ______________ **Liliaceae,** p. 11
18. Ovary inferior.
28. Plants growing in water.
29. Leaves whorled _____ **Hydrocharitaceae** *
29. Leaves basal, or cauline and alternate.

30. Stamens 2, or 6–12, never 3; flowers unisexual; styles not petaloid____________________________**Hydrocharitaceae** *
30. Stamens 3; flowers bisexual; styles petaloid________________**Iridaceae**, p. 150

28. Plants growing on land.
31. Flowers bilaterally symmetrical; stamens 1 or 2_______________**Orchidaceae**, p. 180
31. Flowers radially symmetrical or nearly so; stamens 3 or 6.
32. Stamens 3; styles sometimes petaloid______________________**Iridaceae**, p. 150
32. Stamens 6; styles not petaloid.
33. Leaves reduced to scales; plants lacking chlorophyll, at most 4 cm tall________________**Burmanniaceae**, p. 178
33. Leaves blade-bearing; plants with chlorophyll, well over 4 cm tall_______________________**Liliaceae**, p. 11

Descriptions and Illustrations

Order Liliales

This order, as recognized in this work, includes several groups recognized by some workers as separate orders, such as Amaryllidales, Dioscoreales, Iridales, Haemodorales, Agavales, and Burmanniales. Current evidence seems to indicate all these groups are similar enough to warrant inclusion in a single order.

In Illinois, this revised Order Liliales is made up of the Liliaceae (including Amaryllidaceae), Smilacaceae, Dioscoreaceae, Iridaceae, and Burmanniaceae. Hutchinson's reorganization of the Amaryllidaceae to include only those species with an umbellate inflorescence does not seem consistent when considered with other features. Likewise, Hutchinson's separation of the arborescent genera into the Agavales does not seem justified.

KEY TO THE FAMILIES OF Liliales IN ILLINOIS

1. Plants climbing or twining (if erect, then usually with a few weak tendrils from the upper axils); leaves net-veined; flowers unisexual.
    2. Inflorescence umbellate; ovary superior; fruit a berry--**Smilacaceae,** p. 128
    2. Inflorescence glomerulate or paniculate; ovary inferior; fruit a capsule----------------------------**Dioscoreaceae,** p. 146
1. Plants erect (tendrils never present); leaves mostly parallel-veined; flowers bisexual or unisexual.
    3. Flowers unisexual.
        4. Leaves net-veined; flowers in umbels-------------------------------------*Smilax ecirrata,* in **Smilacaceae,** p. 145
        4. Leaves parallel-veined; flowers in racemes or panicles---**Liliaceae,** p. 11
    3. Flowers bisexual.
        5. Ovary superior--------------------------**Liliaceae,** p. 11
        5. Ovary inferior.
            6. Stamens 3; styles sometimes petaloid---**Iridaceae,** p. 150
            6. Stamens 6(–4); styles not petaloid.
                7. Leaves reduced to scales; plants lacking chlorophyll, at most 4 cm tall--------------**Burmanniaceae,** p. 178

7. Leaves blade-bearing; plants with chlorophyll, well over 4 cm tall_____________________**Liliaceae,** p. 11

LILIACEÆ—LILY FAMILY

Erect perennial herbs, occasionally becoming subligneous, from bulbs, rhizomes, corms, tubers, or thick fibrous roots; inflorescence various; flowers radially symmetrical, perfect (unisexual in a few genera), perianth parts (4)6, free or united at base or into a tube, generally uniform in color; stamens (4)6; ovary superior or inferior; fruit a few- to many-seeded capsule or berry.

The Liliaceae, as constituted here, includes the traditional families Liliaceae and Amaryllidaceae. In conventional systems, the Amaryllidaceae differ from the Liliaceae in the inferior position of the ovary. Hutchinson, on the other hand, has called for the transfer of the liliaceous genera with an umbellate inflorescence (such as *Allium* and *Nothoscordum*) to the Amaryllidaceae. He also proposes to segregate the arborescent genera (such as *Yucca* and *Agave*) into the Agavaceae. Likewise, *Trillium,* with a differentiated perianth and whorled leaves, is considered by some to constitute a separate family, the Trilliaceae. All of these family segregations are rejected here because of the artificial characters used by previous workers in trying to segregate them.

In fact, the differences between the conventional Liliaceae and Amaryllidaceae are so vague that these two families are combined in this work.

This large and showy family ranks behind the Poaceae and Cyperaceae in number of species within the monocots in Illinois.

KEY TO THE GENERA OF Liliaceae OF ILLINOIS

1. Ovary superior.
   2. Leaves evergreen, rigid; stems woody_____________25. *Yucca*
   2. Leaves deciduous, mostly not rigid; stems herbaceous.
      3. Flowers borne in umbels; leaves and bulbs usually with a strong odor of onion (except *Nothoscordum* and *Medeola*)
         4. Leaves in two whorls below the flowers_____23. *Medeola*
         4. Leaves never whorled, usually all basal.
            5. Bulb with a strong odor of garlic or onion; ovary 3-celled, with 1–2 ovules per cell________21. *Allium*
            5. Bulb lacking an odor of garlic or onion; ovary 3-celled, with 6–10 ovules per cell_________22. *Nothoscordum*

3. Flowers borne variously, but not in umbels.
   6. Leaves in one or two or several whorls below the flowers.
      7. Leaves net-veined; plants usually less than 50 cm tall; flower solitary________________________24. *Trillium*
      7. Leaves parallel-veined; plants more than 50 cm tall; flowers usually several_________________7. *Lilium*
   6. Leaves alternate, opposite, or basal, not whorled.
      8. Perianth parts 4; stamens 4; style 2-lobed; leaves 1–3, ovate, cordate__________________20. *Maianthemum*
      8. Perianth parts 6; stamens 6; style 1, simple or 3-cleft, or styles 3, distinct; leaves 1-several, narrower, if ovate, then not cordate.
         9. Flower solitary, one per plant, borne on a leafless scape; leaves 1–2, basal_________14. *Erythronium*
         9. Flowers numerous, or if borne singly, then the stem leafy.
            10. Flowers axillary from the cauline leaves.
                11. Flowers yellow, at first terminal, at length appearing axillary; style deeply 3-cleft; fruit a capsule_________________15. *Uvularia*
                11. Flowers white or greenish, axillary from the first; style unbranched; fruit a berry.
                    12. Flowers perfect, more than 9 mm long; leaves broad; fruit blue or black_____ __________________16. *Polygonatum*
                    12. Flowers unisexual, less than 5 mm long; leaves reduced to minute scales, in the axils of which are filiform branchlets (commonly mistaken for leaves); fruit red_________________19. *Asparagus*
            10. Flowers terminal.
                13. Flower borne singly.
                    14. Style deeply 3-cleft; flowers at most 4 cm long, basically yellowish; plants rhizomatous, to 50 cm tall__________ _____________________15. *Uvularia*
                    14. Style undivided; flowers at least 5 cm long, basically orange; plants bulbous, almost always well over 50 cm tall___ ________________________7. *Lilium*
                13. Flowers in spikes, racemes, panicles, or irregular clusters.

15. Flowers at least 5 cm long, irregularly arranged.
    16. Plants bulbous; stems leafy______ ____________________7. *Lilium*
    16. Plants with fleshy roots, tubers, or rhizomes; stems nearly leafless.
        17. Flowers yellow or orange; leaves linear, to 2 cm broad ____________8. *Hemerocallis*
        17. Flowers lilac or whitish; leaves lanceolate, at least 3 cm broad. __________________9. *Hosta*
15. Flowers less than 5 cm long, arranged in spikes, racemes, or panicles.
    18. Leaves cauline, elliptic to ovate, not grasslike; style 1; fruit a berry.
        19. Perianth parts united into a tube; raceme one-sided______ ____________18. *Convallaria*
        19. Perianth parts free, except at base; raceme or panicle not one-sided______17. *Smilacina*
    18. Leaves basal, if cauline, then long and narrow (grass-like); styles 1 or 3; fruit a capsule.
        20. Style 1; plants bulbous and with a racemose inflorescence.
            21. Perianth parts basically white.
                22. Perianth parts free, with a green stripe down the back, without scales__________ ___10. *Ornithogalum*
                22. Perianth parts united into a long tube, without a green stripe down the back, scaly _________11. *Aletris*
            21. Perianth parts lavender or purple or blue (occasionally white in *Camassia*).

23. Perianth parts united nearly to tip; raceme dense; flowers deep blue or purple---------------12. *Muscari*

23. Parianth parts essentially free; raceme less dense; flowers lavender or pale blue (rarely white)-----------------13. *Camassia*

20. Styles 3; plants rhizomatous or, if bulbous, then with a paniculate inflorescence.

24. Axis of inflorescence glabrous; plants bulbous or rhizomatous.

25. Plants dioecious, rhizomatous; inflorescence a spike-like raceme; capsule loculicidal-----------------6. *Chamaelirium*

25. Plants monoecious or the flowers perfect; plants bulbous; inflorescence a panicle; capsule septicidal.

26. Leaves principally basal; perianth parts with a large gland just below the middle----------4. *Zigadenus*

26. Leaves principally cauline; perianth eglandular---5. *Stenanthium*

24. Axis of inflorescence with a sticky exudate (glutinous), or pubescent; plants rhizomatous.

27. Nearly all leaves basal; inflorescence densely racemose; all flowers perfect---------------1. *Tofieldia*
27. Leaves cauline; inflorescence paniculate; some flowers unisexual.
    28. Perianth parts with 2 glands near the base, clawed; flowers yellow or greenish--3. *Melanthium*
    28. Perianth parts eglandular and without claws; flowers maroon------2. *Veratrum*

1. Ovary inferior.
    29. Flowers campanulate, with the white perianth segments tipped with green----------------------------------26. *Leucojum*
    29. Flowers not campanulate, the perianth parts not green-tipped.
        30. Flowers with a corona, often borne singly or, if in an umbel of 2–6, white.
            31. Flowers recurved; perianth tube (not corona) up to 2 (–4) cm long; stamens included within the corona; ovules several per cell of ovary; bract one--27. *Narcissus*
            31. Flowers ascending; perianth tube (not corona) 5–10 cm long; stamens exserted from the corona; ovules 2 per cell of ovary; bracts 2-several---28. *Hymenocallis*
        30. Flowers without a corona, borne in spikes or, if in an umbel of 2–6, yellow.
            32. Perianth parts united to form a tube, greenish-yellow; leaves large, glabrous, fleshy---------29. *Polianthes*
            32. Perianth parts free, bright yellow; leaves small, pubescent, not fleshy-----------------------30. *Hypoxis*

1. *Tofieldia* HUDS.–False Asphodel

Inflorescence densely racemose; flowers perfect; perianth parts 6, free, glandless; styles 3; capsule septicidal; rhizomatous plants with narrow, basal leaves.

Only the following species occurs in Illinois.

1. Tofieldia glutinosa (Michx.) Pers. Syn. 1:399. 1805. *Fig. 1.*

Narthecium glutinosum Michx. Fl. Bor. Am. 1:210. 1803.

Triantha glutinosa (Michx.) Baker, Journ. Linn. Soc. 17:490. 1879.

Rhizomatous plant to 50 cm tall; leaves several, to 20 cm long, to 8 mm wide, 2-ranked; flowering stalk with a single bract-like leaf, the stalk glutinous; raceme to 5 cm long, densely flowered; flowers white, in groups of 2 or 3, the pedicels 3–6 mm long; perianth parts petaloid, oblanceolate, 3.5–4.5 mm long; capsule ovoid, 4–8 mm long, yellowish or reddish; seeds 1.0–1.5 mm long, with a slender "tail" at each end.

COMMON NAME: False Asphodel.

HABITAT: Bogs.

RANGE: Newfoundland to Alaska, south to California, Illinois, and New York; in the mountains from West Virginia to Georgia.

ILLINOIS DISTRIBUTION: Rare; restricted to the extreme northeastern counties.

The glutinous flower stalk distinguishes this species from all others in the family.

The flowers appear in late June and the first part of July.

For a discussion of variation within *Tofieldia glutinosa*, see Hitchcock (1944).

2. *Veratrum* L.–False Hellebore

Inflorescence paniculate; flowers perfect and staminate; perianth parts 6, free, glandless, without claws; styles 3; capsule septicidal; rhizomatous plants with broad leaves and maroon flowers (in the Illinois species).

Only the following species occurs in Illinois.

1. *Tofieldia glutinosa* (False Asphodel). *a*. Habit, X⅜. *b*. Flower, X2½. *c*. Capsule, X1½.

1. **Veratrum woodii** Robbins in Wood, Class-Book 557. 1855. *Fig. 2.*

Rhizomatous plant to 1.5 m tall; stems leafy, slender; lower leaves elliptic, the upper linear-elliptic, becoming reduced and bract-like; panicle to 50 cm long, the lower flowers staminate, the upper flowers perfect; pedicels 3–6 mm long; perianth parts 6–9 mm long, maroon, obtuse to subacute at apex, narrowed but not clawed at the base; capsule ovoid, 20–25 mm long.

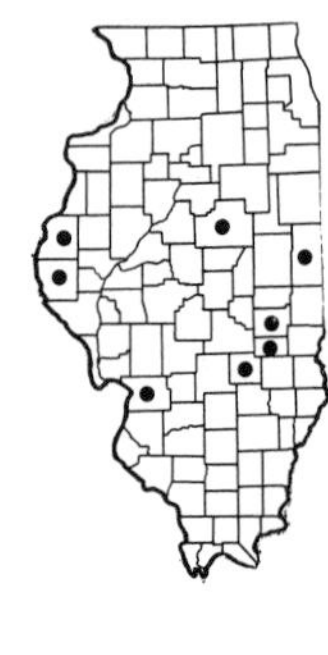

COMMON NAME: False Hellebore.

HABITAT: Rich, moist woodlands.

RANGE: Ohio, Indiana, Illinois, Iowa, Missouri, and Oklahoma.

ILLINOIS DISTRIBUTION: Rare; known from eight counties through the central part of the state.

This species has one of the most limited overall ranges of any species occurring in Illinois. It is known from only six states, and is apparently not abundant in any state.

The flowers appear in July and August.

The inflorescence, as in *Melanthium,* contains both staminate and perfect flowers, but the perianth parts, unlike *Melanthium,* are glandless. There have been attempts by some to combine *Veratrum* and *Melanthium.*

3. *Melanthium* L. – Bunch-flower

Inflorescence paniculate; flowers perfect and staminate; perianth parts 6, free, each bearing two glands near the base, clawed; styles 3, distinct; capsule septicidal; rhizomatous plants with yellowish or greenish flowers.

Only the following species occurs in Illinois.

1. **Melanthium virginicum** L. Sp. Pl. 339. 1753. *Fig. 3.*

Rhizomatous plant to 1.5 m tall, the stems hairy above; leaves basal, linear, acuminate, to 30 cm long, to 2.5 cm broad; panicle 25–35 cm long, the axis pubescent, the lower flowers perfect, the upper flowers staminate; perianth parts 5–12 mm long, yellowish or greenish, elliptic to obovate, subacute, scurfy on the back, with a distinct claw, with 2 dark glands near the base of each; capsule ellipsoid, 12–16 mm long; seeds obovoid, whitish, 5–8 mm long.

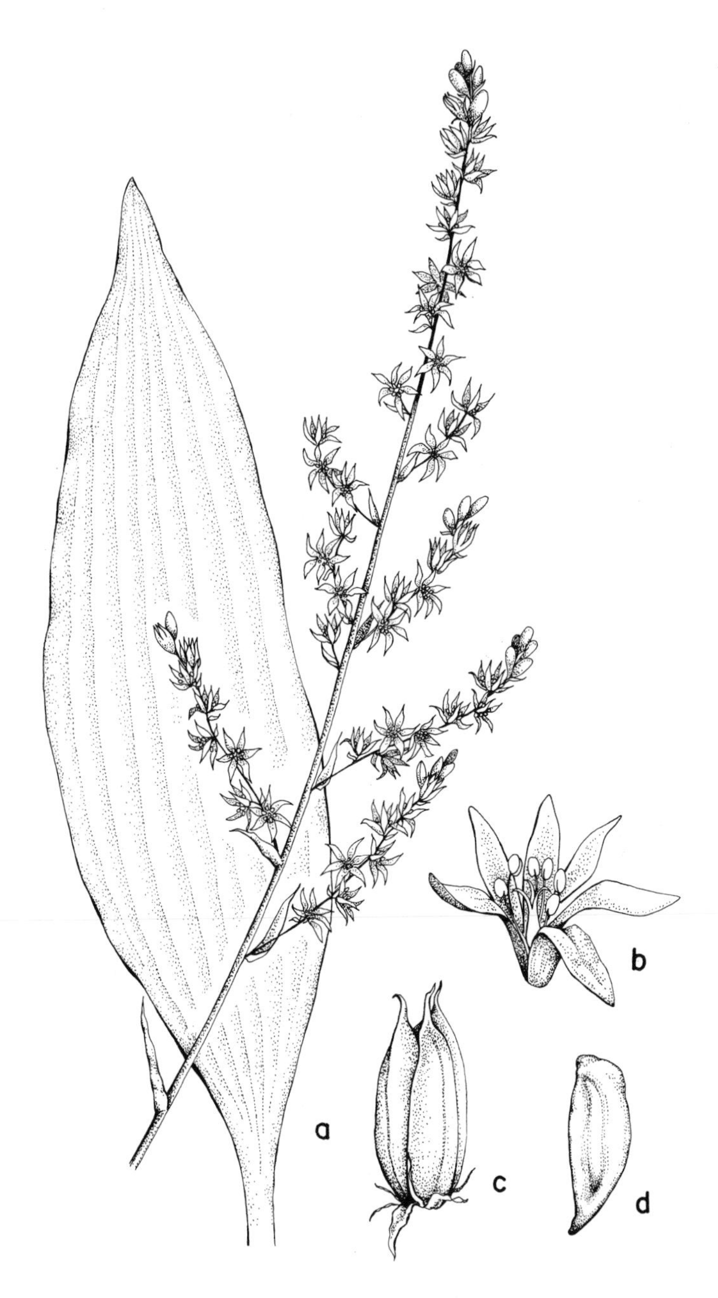

2. *Veratrum woodii* (False Hellebore). *a.* Inflorescence and leaf, X¼. *b.* Flower, X1½. *c.* Capsule, X¾. *d.* Seed, X½.

3. *Melanthium virginicum* (Bunch-flower). *a.* Inflorescence and leaf, X⅛. *b.* Staminate flower, X1¼. *c.* Perfect flower, X1¼. *d.* Capsule, X1¼. *e.* Seed, X2½.

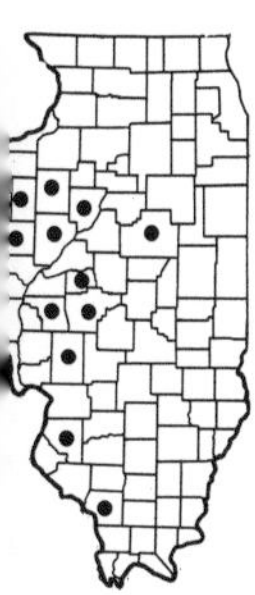

COMMON NAME: Bunch-flower.
HABITAT: Meadows; wet prairies.
RANGE: New York to Minnesota, south to Texas and Florida.
ILLINOIS DISTRIBUTION: Occasional in the west-central counties; rare in the southwestern counties; absent from all other parts of the state.
This handsome species flowers during late June and July.

The presence of two dark glands near the base of each part of the perianth is diagnostic for this species.

4. *Zigadenus* MICHX. – Camass

Inflorescence paniculate; flowers perfect; perianth parts 6, free, each bearing a single gland (in the Illinois species) just below the middle, without a claw; styles 3, distinct; capsule septicidal; bulbous plants with whitish flowers.

Only the following species occurs in Illinois.

1. **Zigadenus glaucus** Nutt. Journ. Acad. Nat. Sci. Phila. 7:56. 1834. *Fig. 4.*

Bulbous plant to 75 cm tall, the stems glabrous; leaves mostly basal, linear, coriaceous, obtuse to subacute, to 35 cm long, to 1.2 cm broad; cauline leaves few, much reduced and bract-like; panicle 25–40 cm long, the axis glabrous, bracteate; perianth parts 8–14 mm long, whitish or greenish suffused with purple on the back, elliptic to obovate, with a dark, bilobed gland near the base of each, without a claw; capsule ovoid, 10–15 mm long; seeds oblanceoloid, 3.0–4.5 mm long.

COMMON NAME: White Camass.
HABITAT: Limestone cliffs or low areas near rivers.
RANGE: Quebec to Minnesota, south to Illinois and New York; in the mountains from West Virginia to North Carolina.
ILLINOIS DISTRIBUTION: Rare; known only from JoDaviess and Kane counties.
The white camass flowers from mid-July until the first week in September.

The specific epithet is derived from the very glaucous bulb. Each perianth part bears a single bilobed gland.

4. *Zigadenus glaucus* (White Camass). *a.* Inflorescence and leaves, X⅛. *b.* Flower, X1¼. *c.* Capsule, X2.

The original locality of this species in Illinois is in *Carex* marshes along the Fox River in Elgin, adjacent to Trout Park. The colony still survives there. The JoDaviess station, on the other hand, is on limestone cliffs.

5. *Stenanthium* GRAY – Grass-leaved Lily

Inflorescence paniculate; flowers perfect or unisexual; perianth parts 6, free, glandless, without a claw; styles 3, distinct; capsule septicidal; slightly bulbous plants with greenish or whitish flowers.

Only the following species comprises the genus.

1. Stenanthium gramineum (Ker) Morong, Mem. Torrey Club 5:110. 1894. *Fig. 5.*

Helonias graminea Ker, Bot. Mag. pl. 1599. 1813.
Veratrum angustifolium Pursh, Fl. Am. Sept. 242. 1814.
Stenanthium angustifolium (Pursh) Kunth, Enum. 4:190. 1843.
Stenanthium robustum Wats. Proc. Am. Acad. 14:278. 1879.
Stenanthium gramineum var. *typicum* Fern. Rhodora 48:151. 1946.

Slightly bulbous plant to 1.5 (–1.7) m tall, the stems glabrous; leaves cauline, lanceolate, acute, to 40 cm long, to 3 cm broad, strongly ascending, the upper becoming much reduced and bract-like; panicle 40–75 cm long, the lower flowers mostly staminate, the upper perfect, the axis glabrous; perianth parts 5–10 mm long, whitish or greenish, lanceolate, acute to acuminate, glandless, without a claw; capsule ovoid, 8–14 mm long; seeds lanceoloid, 3–7 mm long.

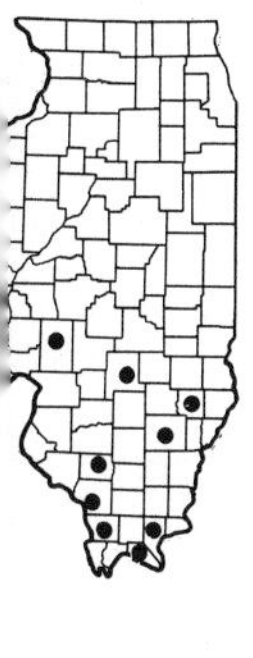

COMMON NAME: Grass-leaved Lily.

HABITAT: Moist woodlands, particularly around woodland ponds and along streams.

RANGE: Pennsylvania to Missouri, south to Texas and Florida.

ILLINOIS DISTRIBUTION: Not common; confined to the southern two-fifths of Illinois.

The flowering time for this species in Illinois is from mid-June to mid-August.

Fernald (1946) has discussed the status of *Stenanthium gramineum*, *S. robustum*, and *S. angustifolium*. There seems to be no justification for maintaining as distinct more robust speci-

5. *Stenanthium gramineum* (Grass-leaved Lily). *a.* Inflorescence and leaves, X⅛. *b.* Flower, X2½. *c.* Capsules, X½. *d.* Seed, X2.

mens (*S. robustum*) or specimens with narrow leaves (*S. angustifolium*).

6. *Chamaelirium* WILLD. – Devil's Bit

Inflorescence a terminal, spike-like raceme; flowers unisexual; perianth parts 6, free, glandless, without a claw; styles 3, distinct; capsule loculicidal; rhizomatous, dioecious plants with whitish or greenish flowers.

Only the following species comprises the genus.

1. **Chamaelirium luteum** (L.) Gray, Man. Bot. 503. 1848. *Fig. 6.*

Veratrum luteum L. Sp. Pl. 1044. 1753.
Helonias dioica Pursh, Fl. Am. Sept. 1:243. 1814.
Chamaelirium obovale Small, Torreya 1:108. 1901.

Rhizomatous, dioecious plants; staminate plants to 70 cm tall, glabrous; pistillate plants usually nearly twice as tall, glabrous; leaves nearly all basal, obovate, the lower to 15 cm broad; cauline leaves few, becoming much reduced and bract-like; inflorescence spike-like, racemose; perianth parts 2.5–3.0 mm long, whitish or greenish, becoming yellowish after flowering, linear, subacute or obtuse, glandless, without a claw; capsule ellipsoid, to nearly 15 mm long; seeds narrowly oblongoid, 3–5 mm long.

COMMON NAME: Fairy Wand.
HABITAT: Low, wooded hillsides.
RANGE: Massachusetts to Ontario, south to Arkansas and Florida.
ILLINOIS DISTRIBUTION: Rare; known only from Hardin, Massac, and Pope counties. At one of the Pope County stations, near Massac Tower, over one hundred plants occur. A specimen in the Gray Herbarium from Abingdon exists, but this almost certainly is not Knox County, Illinois, where Abingdon is located.

The fairy wand flowers from mid-May to late June. It was first found in Illinois in 1932. The dioecious nature of the plant, along with the spike-like raceme and the obovate basal leaves, distinguishes this plant from similar genera such as *Stenanthium, Zigadenus, Veratrum, Melanthium,* and *Tofieldia.* In addition, the above genera have septicidal capsules, while the capsule in *Chamaelirium* is loculicidal.

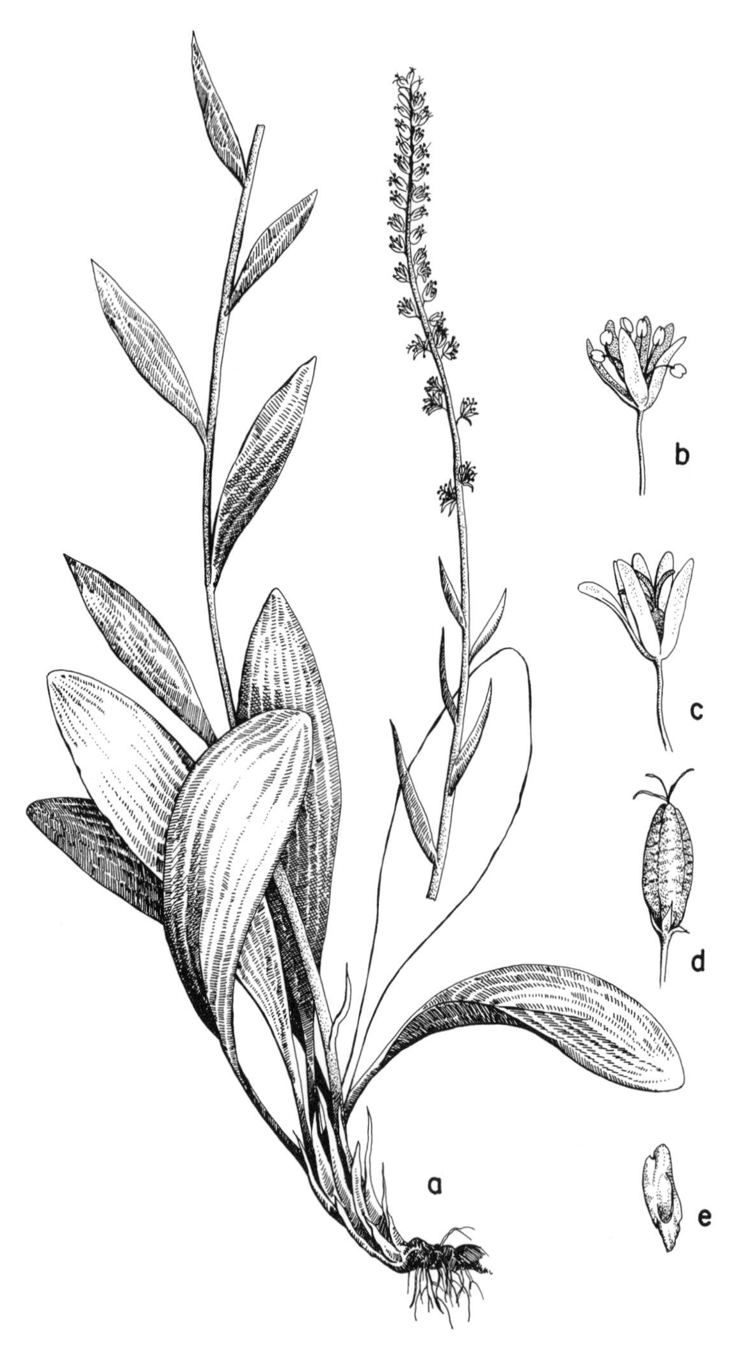

6. *Chamaelirium luteum* (Fairy Wand). *a.* Inflorescence and leaves, X⅛. *b.* Staminate flower, X2½. *c.* Pistillate flower, X2½. *d.* Capsule, X¾. *e.* Seed, X1½.

7. *Lilium* L. – Lily

Inflorescence terminal, the flowers irregularly arranged; flowers perfect; perianth parts 6, united below into a campanulate or funnelform tube; style 1; capsule loculicidal; tall, showy-flowered plants from a bulbous base, and with cauline leaves.

Several lilies are grown in cultivation in Illinois. Of those, only *Lilium lancifolium* escapes, and this not commonly. In addition, three native species occur in Illinois. *Lilium catesbaei* Walt., a southeastern species questionably reported from Illinois by Fernald (1950), apparently does not occur in Illinois.

KEY TO THE TAXA OF Lilium IN ILLINOIS

1. All the leaves alternate, the upper with bulblets in the axils; stems scabrous ------------------------------------ 1. *L. lancifolium*
1. Some of the leaves borne in whorls; bulblets absent; stems glabrous.
   2. Flowers erect; perianth parts with claws; only the uppermost group of leaves in a whorl; plants less than 1 m tall ------------------------------------ 2. *L. philadelphicum* var. *andinum*
   2. Flowers nodding; perianth parts without claws; several whorls of leaves borne on the stem; plants generally over 1 m tall.
      3. Bulbs yellow; margins and nerves of leaves roughened; midvein of outer perianth parts rounded on the back; anthers 8–15 mm long; filament attached 1–2 mm from end of anther ------------------------------------ 3. *L. michiganense*
      3. Bulbs white; margins and nerves of leaves smooth; midvein of outer perianth parts sharply ridged on the back; anthers 17–25 mm long; filament attached 4–8 mm from end of anther ------------------------------------ 4. *L. superbum*

1. Lilium lancifolium Thunb. Trans. Linn. Soc. 2:333. 1794. *Fig. 7.*

Lilium tigrinum Andr. Bot. Rep. 9:errata. 1809.

Lilium tigrinum Ker, Curtis' Bot. Mag. 31:plate 1237. 1809.

Stem over 1 m tall, scabrous, purplish; leaves alternate, narrowly lanceolate, the lower to 15 cm long, the upper shorter and with bulblets in the axils; flowers several, nodding, the pedicels pubescent, 6–12 cm long, bracteate; perianth parts orange or orange-red, spotted with dark purple, to 10 cm long, strongly recurved, the midveins pubescent at base; 2n = 36 (Sato, 1932).

7. *Lilium lancifolium* (Tiger Lily). X⅛.

COMMON NAME: Tiger Lily.
HABITAT: Escaped from cultivation into waste ground or along roads.
RANGE: Native of eastern Asia.
ILLINOIS DISTRIBUTION: Scattered; not commonly escaped from cultivation, and apparently only collected once.
The flowers of the tiger lily appear in July and August. Capsules are rarely produced in the cultivated forms. The roughened stem and the alternate leaves distinguish this species from all other Illinois lilies.

Ingram (1968) has given reason why *Lilium lancifolium* should be the correct binomial for the tiger lily, rather than *L. tigrinum.*

2. **Lilium philadelphicum** L. var. **andinum** (Nutt.) Ker, Bot. Reg. 7:pl. 594. 1822. *Fig. 8.*

Lilium umbellatum Pursh, Fl. Am. Sept. 229. 1814.
Lilium andinum Nutt. Gen. N. Am. Pl. 1:221. 1818.

Stem never over 1 m tall, glabrous; uppermost leaves whorled, the lower alternate, narrowly lanceolate, to 10 cm long, to 1 cm broad, without axillary bulblets; flowers 1–5, erect, the pedicels glabrous, stout; perianth parts ascending, orange to orange-red to red, purple-spotted within, lance-ovate to ovate, tapering to a distinct claw at the base; capsule 4–7 cm long; $2n = 24$ (Sansome & LaCour, 1934).

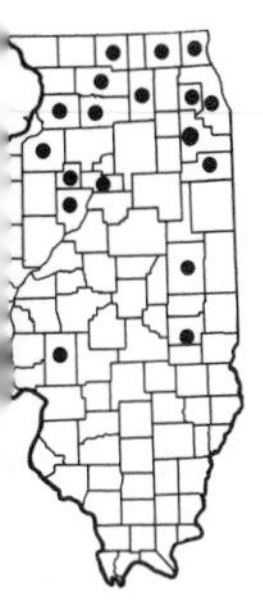

COMMON NAME: Wood Lily; Western Lily.
HABITAT: Dry woodlands.
RANGE: Quebec to British Columbia, south to New Mexico and Kentucky.
ILLINOIS DISTRIBUTION: Occasional; restricted to the upper half of the state.
June and July are the months when this species flowers. This variety differs from the more eastern var. *philadelphicum* by its shorter capsules and narrower leaves, only the uppermost of which are borne in whorls. Authors who choose to recognize this taxon as a species must call it *L. umbellatum.*

The western lily is the smallest in stature of the native lilies in Illinois, and the only taxon in which the flowers are borne erect.

8. *Lilium philadelphicum* var. *andinum* (Wood Lily). *a.* Inflorescence and upper leaves, X¼. *b.* Capsule, X⅛. *c.* Seed, X½.

3. **Lilium michiganense** Farw. Bull. Torrey Club 42:353. 1915. *Fig. 9.*

Bulbs yellow; stem to nearly 2 m tall, often shorter, glabrous; some or all of the leaves whorled, lanceolate, roughened along the veins and margins, to 10 (–12) cm long, to 2 cm broad, without axillary bulblets; flowers 1-several, nodding, the pedicels glabrous; perianth parts strongly recurved, to 8 cm long (usually shorter), orange or orange-red, spotted with dark purple, the outer three with a rounded ridge on the back; anthers 8–15 mm long; filaments attached 1–2 mm from end of anther; $2n = 24$ (Stewart, 1947).

COMMON NAME: Turk's-cap Lily.

HABITAT: Moist woodlands and prairies.

RANGE: Ontario to Manitoba, south to Arkansas and Tennessee.

ILLINOIS DISTRIBUTION: Rather common throughout the state.

This beautiful wild flower blooms the last week in May until about the middle of July.

This species and the following have been confused by many taxonomists, with both usually being grouped together. Anyone seeing the two species in the field, however, can recognize immediately the specific differences between the two. Deam (1940) and Wherry (1942) have given excellent accounts of these differences. They are summarized in the preceding key.

4. **Lilium superbum** L. Sp. Pl. 434. 1762. *Fig. 10.*

Bulbs white; stem to 2 m tall, glabrous; some or all the leaves whorled, lanceolate to lance-ovate, smooth along the veins and margins, to 10 cm long, to 2 cm broad, without axillary bulblets; flowers 1-few, nodding, the pedicels glabrous; perianth parts strongly recurved, to 10 cm long, orange-red or orange, spotted with dark purple, the outer three with a sharp-angled ridge on the back; anthers 17–25 mm long; filaments attached 4–8 mm from end of anthers.

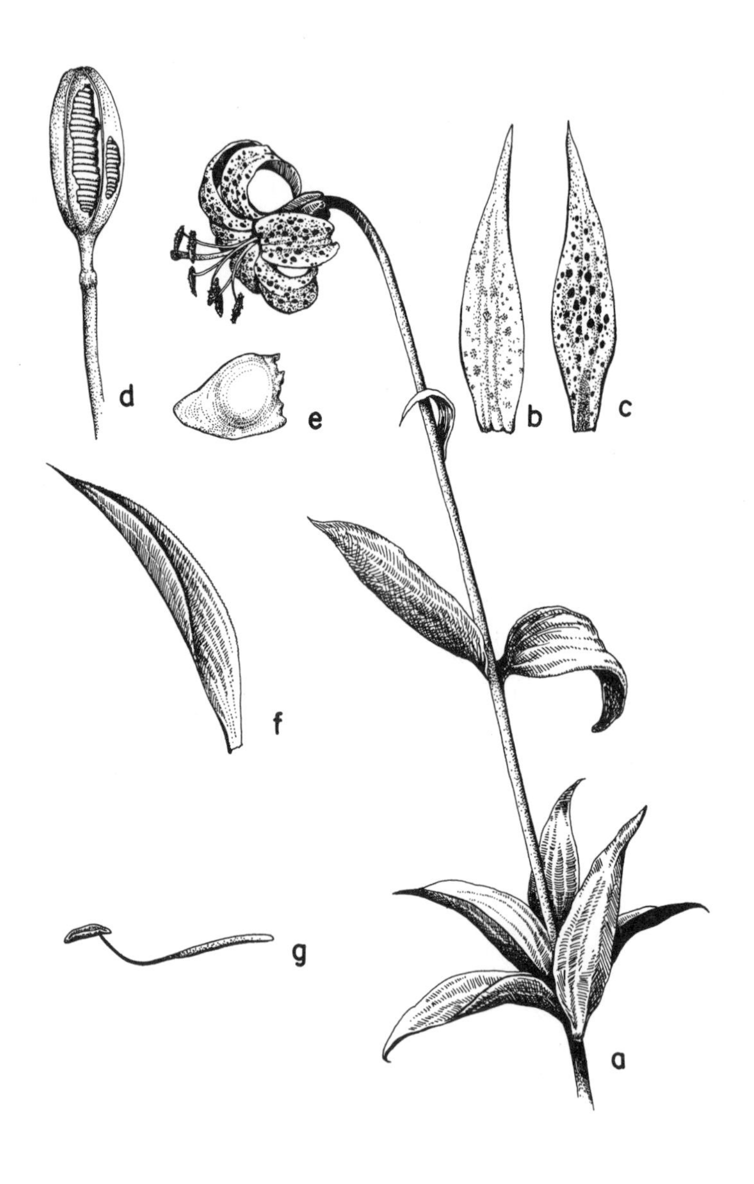

9. *Lilium michiganense* (Turk's-cap Lily). *a.* Inflorescence and upper leaves, X⅛. *b.* Petal, dorsal view, X⅓. *c.* Petal, ventral view, X⅓. *d.* Capsule, X⅛. *e.* Seed, X¼. *f.* Leaf, X⅓. *g.* Stamen, X⅓.

10. *Lilium superbum* (Superb Lily). *a.* Inflorescence and upper leaves, X⅛. *b.* Petal, dorsal view, X⅓. *c.* Petal, ventral view, X⅓. *d.* Leaf, X¼. *e.* Stamen, X⅓.

COMMON NAME: Superb Lily; Turk's-cap Lily.
HABITAT: Low, moist woodland.
RANGE: New Hampshire to southern Illinois, south to Alabama and Georgia.
ILLINOIS DISTRIBUTION: Very rare; known only from a single colony at Lake Murphysboro State Park, Jackson County, and Lusk Creek, Pope County.

This species, which forms a colony of several hundred plants in Jackson County, had not been observed to flower in the wild in Illinois until 1966, as all plants die back before flowering. On July 22, 1966, one specimen reached flowering stage at Lake Murphysboro. For an account of the discovery of this species in Illinois, see Mohlenbrock (1962). A colony of this species also occurs at Lusk Creek, Pope County, but most of these plants die back before flowering, also.

This is undoubtedly one of the most handsome of Illinois wild flowers. The flowers, which appear in early or mid-July, open later than those of *L. michiganense*.

8. *Hemerocallis* L. – Day Lily

Inflorescence terminal, the flowers irregularly arranged; flowers perfect; perianth parts 6, united below into a funnelform tube; style 1; capsule loculicidal; tall, showy-flowered plants from fleshy roots, and with basal leaves.

The common name results from the flowers which last but a single day.

Many cultivated forms are grown in Illinois, but only the following two escape with any frequency.

KEY TO THE SPECIES OF Hemerocallis IN ILLINOIS

1. Flowers orange ---------------------------------- 1. *H. fulva*
1. Flowers yellow -------------------------- 2. *H. lilio-asphodelus*

1. Hemerocallis fulva L. Sp. Pl. 462. 1762. *Fig. 11.*

Stem to 1 m tall (or taller), glabrous; leaves basal, 2-ranked, elongate-linear, to 35 cm long, to 2 cm broad; flowers 3–15, ascending, the pedicels glabrous; perianth parts recurved, 8–15 cm long, obtuse to subacute, orange, the inner three wavy-margined; $2n = 22$ (Stout, 1932).

11. *Hemerocallis fulva* (Day Lily). X⅛.

COMMON NAME: Orange Day Lily.
HABITAT: Roadsides, waste areas.
RANGE: Introduced from Eurasia.
ILLINOIS DISTRIBUTION: Common throughout the state; probably escaped in every county.
This species spreads rapidly by means of its tuberous roots. It flowers from June through August.

2. **Hemerocallis lilio-asphodelus** L. Sp. Pl. 324. 1753. Not illustrated.

Hemerocallis flava L. Sp. Pl. 462. 1762.
Stem to 1 m tall (or taller), glabrous; leaves basal, 2-ranked, elongate-linear, to 35 cm long, to nearly 2 cm broad; flowers 3–12, odorous, ascending, the pedicels glabrous; perianth parts recurved, 8–15 cm long, obtuse to subacute, yellow, the inner three wavy-margined; 2n = 22 (Dark, 1932).

COMMON NAME: Yellow Day Lily.
HABITAT: Roadsides, waste areas.
RANGE: Introduced from Asia.
ILLINOIS DISTRIBUTION: Occasionally escaped throughout the state, but rarely collected.
This species escapes and becomes naturalized much less frequently than the orange day lily. It also flowers from June through August.
Farwell (1928) was the first to point out the illegitimate nature of the binomial *H. flava.*

9. *Hosta* TRATT – Plantain Lily

Inflorescence terminal, the flowers irregularly arranged; flowers perfect; perianth parts 6, united below into a funnelform tube; style 1; capsule loculicidal; rather tall, showy-flowered plants from fibrous roots and rhizomes, with broad basal leaves.

The following cultivated plant rarely escapes cultivation in Illinois.

1. **Hosta lancifolia** (Thunb.) Engl. in Engl. & Prantl, Nat. Pflanz. 2:5. 1887. *Fig. 12.*

Hemerocallis japonica Thunb. Fl. Jap. 142. 1784, non *Hosta japonica* Tratt (1812).

12. *Hosta lancifolia* (Plantain Lily). X5/16.

Hemerocallis lancifolia Thunb. Trans. Linn. Soc. 2:335. 1794.
Hosta japonica (Thunb.) Voss in Vilmorin, Blumengaertnerei 1:1076. 1896.

Stem usually less than 1 m tall, glabrous; leaves basal, broadly lanceolate, to 10 cm long, to 4 cm broad, acute; flowers several, at maturity nodding, the pedicels glabrous; perianth parts recurved at the tips, to 4 cm long, pale lilac or whitish; 2n = 60 (Akemine, 1935; Matsuura & Sutô, 1935).

COMMON NAME: Plantain Lily.
HABITAT: Escaped into disturbed areas.
RANGE: Native to eastern Asia.
ILLINOIS DISTRIBUTION: Rarely escaped; recorded only from Carroll County (Mississippi Palisades State Park, August 21, 1956, *R.H. Mohlenbrock 7214*).

10. *Ornithogalum* L. – Star-of-Bethlehem

Inflorescence racemose; flowers perfect; perianth parts 6, free, with a broad green stripe on the back; style 1; capsule loculicidal; bulbous plants with basal leaves.

Only the following species occurs in Illinois.

1. Ornithogalum umbellatum L. Sp. Pl. 307. 1753. *Fig. 13.*

Bulbs tunicated; leaves linear, up to 15 cm long, to 4 mm broad; scape to 30 cm tall, bracted; flowers 3–7, the pedicels 2–6 cm long, glabrous; perianth parts spreading, broadly lanceolate, acute, 1.5–2.0 cm long, white, with a green stripe on the back; 2n = 27 (Gadella & Kliphuis, 1963); 54 (Matsuura & Sutô, 1935).

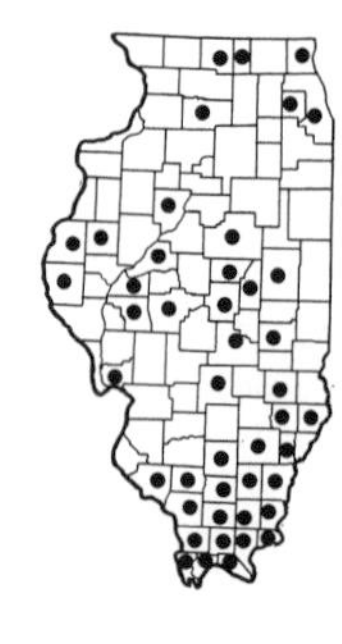

COMMON NAME: Star-of-Bethlehem.
HABITAT: Roadsides; open grassy areas; in woods.
RANGE: Native to Europe; escaped throughout the United States.
ILLINOIS DISTRIBUTION: Common; throughout the state, often forming extensive patches.

This species flowers from mid-April to mid-June. The green stripe down the back of each perianth part readily distinguishes this species. Under favorable conditions, this species may run rampant and com-

13. *Ornithogalum umbellatum* (Star-of-Bethlehem). *a.* Habit, X¼. *b.* Flower, X1.

pletely invade an area. It is not uncommon to find this species escaped into woodlands. It occurs in the nearly virgin forest at Martha's Woods in Pope County.

11. *Aletris* L. – Colic Root

Inflorescence a spike-like raceme; flowers perfect; perianth parts 6, united into a long tube, scaly on the back; style 1; capsule loculicidal; rhizomatous plants with a rosette of basal leaves and reduced leaves on the scape.

Only the following species occurs in Illinois.

1. Aletris farinosa L. Sp. Pl. 319. 1753. *Fig. 14.*

Leaves more or less rigid, narrowly lanceolate, acute, the basal rosette leaves to 20 cm long, to 2 cm broad, the leaves of the scape much reduced and bract-like; scape to 100 cm tall; flowers white, tubular; perianth lobes erect, 2.0–2.5 mm long, minutely scaly on the back; capsule ovoid, 3–5 mm long; seeds 0.5–0.7 mm long.

COMMON NAME: Colic Root.

HABITAT: Moist, sandy prairies; sandy flats.

RANGE: Maine to Minnesota, south to Texas and Florida.

ILLINOIS DISTRIBUTION: Not common; restricted to the northern one-third of the state, particularly in the eastern counties.

The mealy appearance of the perianth parts is characteristic for this species, which flowers as early as June 21 and persists until late August. The ovary is about one-third inferior.

Floyd Swink of the Morton Arboretum, who has an intimate knowledge of the plant associations in the northeastern counties of Illinois, reports that *Aletris farinosa* occurs in moist sandy prairies, being associated with *Asclepias incarnata, Calopogon pulchellus, Eupatorium perfoliatum, Fragaria virginiana, Gentiana crinita, Gerardia purpurea, Juncus canadensis, Krigia biflora, Lespedeza capitata, Linum medium* var. *texanum, Ludwigia alternifolia, Pogonia ophioglossoides, Potentilla simplex, Rhynchospora capitellata,* and *Solidago patula.* It also may occur in sunny, acid, sand flats, associated with *Agrostis hyemalis, Bartonia virginica, Cassia fasciculata, Drosera intermedia, D. rotundifolia, Gaultheria procumbens, Melampyrum lineare* var. *latifolium, Polygala cruciata* var. *aquilonia, Rhexia virginica, Rubus hispidus* var. *obovalis, Viola lanceolata,* and *Xyris torta.* When *Aletris farinosa* grows in rich prairies, it is found with *Andropogon gerardii, Aster azureus, A. ericoides, Dodecatheon media, Lithospermum canescens, Petalostemum purpureum,* and *Sisyrinchium albidum.*

12. *Muscari* MILL. – Grape Hyacinth

Inflorescence densely racemose; flowers perfect (occasionally sterile in some species); perianth parts 6, united except at the very tips; style 1; capsule loculicidal; bulbous plants with narrow, basal leaves.

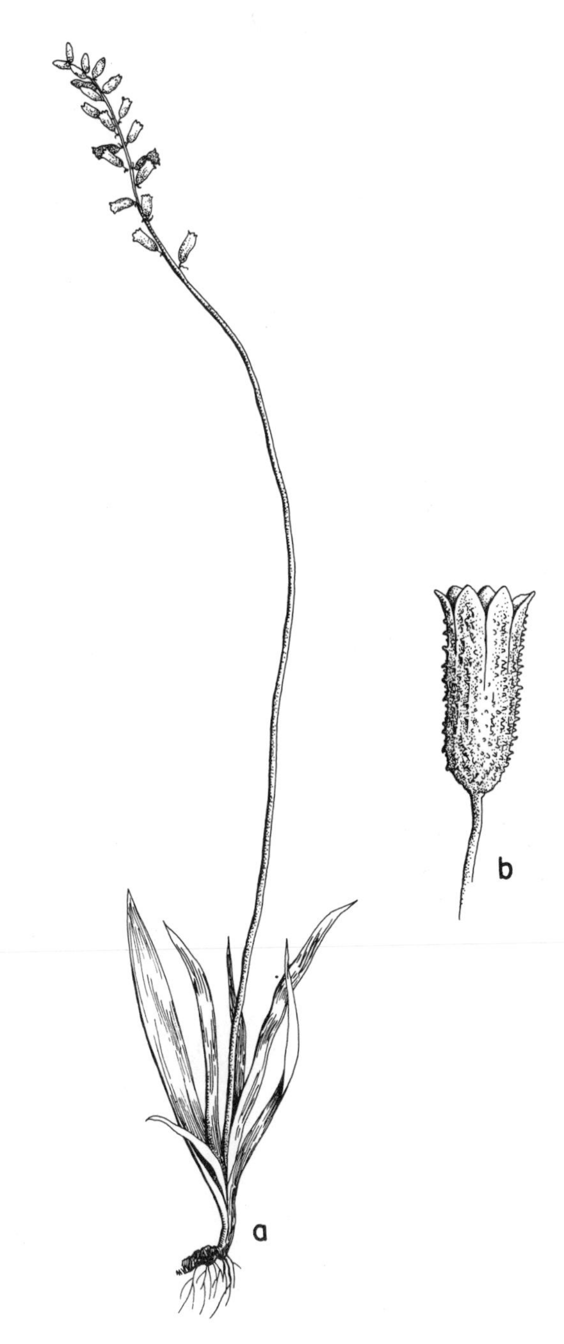

14. *Aletris farinosa* (Colic Root). *a.* Habit, X1. *b.* Flower, X5.

KEY TO THE SPECIES OF Muscari IN ILLINOIS

1. Leaves flat, most or all of them over 3 mm broad.
   2. Raceme up to 6 cm long; perianth to 6 mm long.
      3. All flowers fertile, purple or blue; perianth 3.5–5.0 mm long -- 1. *M. botryoides*
      3. Fertile flowers deep violet with white teeth; sterile flowers pale blue; perianth 5–6 mm long-------- 2. *M. armeniacum*
   2. Raceme at least 9 cm long; perianth 9–11 mm long---------- -- 3. *M. comosum*
1. Leaves terete, not more than 3 mm broad-------- 4. *M. atlanticum*

1. **Muscari botryoides** (L.) Mill. Gard. Dict. ed. 8:1. 1768. *Fig. 15.*

Hyacinthus botryoides L. Sp. Pl. 318. 1753.

Bulbs tunicated; leaves basal, linear to linear-lanceolate, flat, acute or subacute, to 25 cm long, 3–8 mm broad; scapes 8–20 cm long; raceme up to 6 cm long; flowers purple or blue, cup-shaped; perianth 3.5–5.0 mm long, glabrous; capsule subgloboid, 4.5 mm in diameter; seeds rounded, rugulose, 0.5–0.8 mm long; 2n = 36 (Delaunay, 1926); n = 24 (Matsuura & Sutô, 1935).

COMMON NAME: Grape Hyacinth.

HABITAT: Fields, particularly grassy places.

RANGE: Native of Europe; escaped from cultivation nearly throughout the United States.

ILLINOIS DISTRIBUTION: Occasional.

April is the time of flowering for this species. Perianth parts are rather variable in size, as is the height of the scape.

The flat blades of this species distinguish it from *M. atlanticum,* while the shorter racemes and smaller perianth separate it from *M. comosum.*

Muscari botryoides is frequently grown as an early flowering ornamental. Where it has been found as an escape, it often occurs in great abundance.

2. **Muscari armeniacum** Leicht. ex. Baker, Gard. Chron. 1:798. 1878. *Fig. 16.*

Bulbs tunicated; leaves basal, linear to linear-lanceolate, flat, acute or subacute, 6–10 mm broad; scapes 10–22 cm tall,

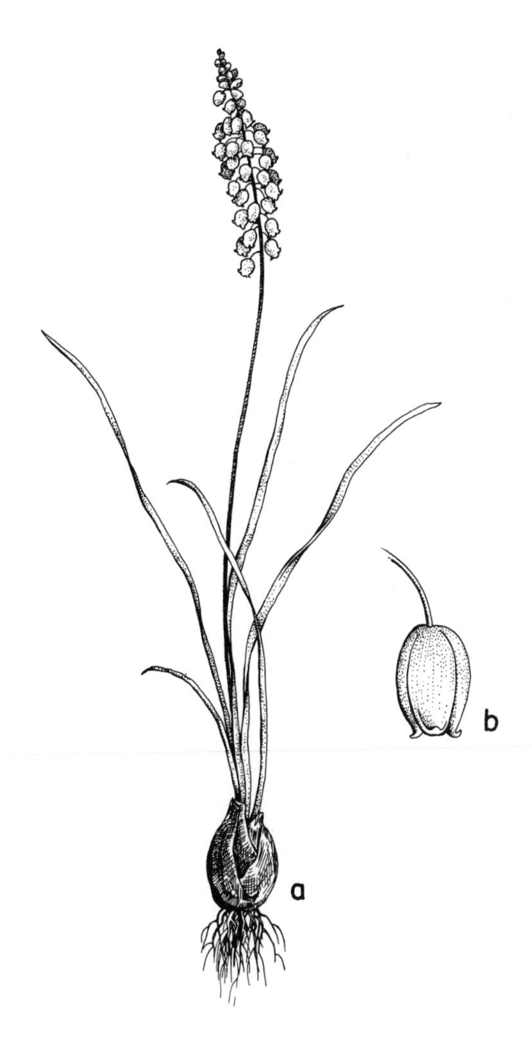

15. *Muscari botryoides* (Grape Hyacinth). *a.* Habit, X¼. *b.* Flower, X1¾.

16. *Muscari armeniacum* (Heavenly Blue). *a.* Habit, X¼. *b.* Flower, X2.

shorter than the leaves; raceme many-flowered, the uppermost flowers sterile, pale blue, the lowermost flowers fertile, deep violet, with erect, white teeth; perianth 5–6 mm long, glabrous; fruit and seeds not observed.

COMMON NAME: Heavenly Blue.
HABITAT: Abandoned cultivated area.
RANGE: Native of southwestern Asia.
ILLINOIS DISTRIBUTION: Escaped from cultivation in Illinois in Piatt County.

This handsome species differs from other members of the genus in Illinois by possessing both sterile and fertile flowers. These flowers are of strikingly different colors.

3. **Muscari comosum** (L.) Mill. Gard. Dict. ed. 8:2. 1768. *Fig. 17.*

Hyacinthus comosus Sp. Pl. 318. 1753.

Bulbs tunicated; leaves 3–4, basal, linear to linear-lanceolate, flat, acute or subacute, to 45 cm long, to 2.4 cm broad; scapes to 30 cm long; raceme 9–15 cm long; upper flowers bluish, sterile, lower flowers olive; perianth 9–11 mm long, glabrous; capsule to 12 mm in diameter; $2n = 18$ (Delaunay, 1926).

COMMON NAME: Grape Hyacinth.
HABITAT: Adventive along edge of field.
RANGE: Native of southern Europe and Asia; frequently escaped in the United States.
ILLINOIS DISTRIBUTION: Jackson County: along road to Carbondale Reservoir, *R. Mohlenbrock s.n.*

This species rarely escapes from cultivation, but a cluster of specimens was discovered along a road in Jackson County in 1962 and recollected from the same location in 1963. The species is much more robust than the other species of *Muscari* escaped in Illinois.

4. **Muscari atlanticum** Boiss. & Reut. Pugill. Pl. Nov. 114. 1852. *Fig. 18.*

Bulbs tunicated; leaves 5–6, basal, linear, nearly terete but channeled on one side, acute or subacute, to 25 cm long, less than 3 mm broad; scapes 12–28 cm long; flowers blue, cup-

17. *Muscari comosum* (Grape Hyacinth). *a.* Habit, X¼. *b.* Flower, X¾. *c.* Fruiting cluster, X¼.

18. *Muscari atlanticum* (Blue Bottle.) *a*. Habit, X5⁄16. *b*. Flower, X2¼.

shaped; perianth 4–6 mm long, glabrous; capsule subgloboid, 4–6 mm in diameter; seeds rounded, 0.7–1.0 mm long; 2n = 54 (Wunderlich, 1937).

COMMON NAME: Blue Bottle.
HABITAT: Waste ground; in fields.
RANGE: Native of Europe; escaped from cultivation in the eastern half of the United States.
ILLINOIS DISTRIBUTION: Occasional.

This species blooms from April to mid-May. It is somewhat larger in most respects than *M. botryoides.* Although this species is usually referred to as *Muscari racemosum* (L.) Mill., Miller did not base this combination on Linnaeus's *Hyacinthus racemosus.* Miller's binominal applies to a completely different plant, the musk hyacinth.

13. *Camassia* LINDL. – Wild Hyacinth

Inflorescence racemose; flowers perfect; perianth parts 6, essentially free; style 1; capsule loculicidal; bulbous plants with narrow, basal leaves.

The narrow basal leaves, perfect flowers, and racemose inflorescence distinguish this genus from all but *Muscari.* From this latter genus it differs by its free perianth segments.

KEY TO THE SPECIES OF Camassia IN ILLINOIS

1. Scape with 0–2 (–3) deciduous bracts; inflorescence at anthesis 3–5 cm broad; capsule as broad as long; perianth parts white to pale blue to pale lilac; plants beginning to flower in early April---1. *C. scilloides*
1. Scape with 3–24 persistent bracts; inflorescence at anthesis 2–3 (–3.5) cm broad; capsule longer than broad; perianth parts deep lavender to pale purple; plants beginning to flower in May--2. *C. angusta*

1. Camassia scilloides (Raf.) Cory, Rhodora 38:405. 1936. *Fig.* 19.

Phalangium esculentum Nutt. ex Ker, Bot. Mag. 38:pl. 1574. 1813, in synon.
Scilla esculenta Ker, Bot. Mag. pl. 1754. 1813.
Cyanotris scilloides Raf. Am. Month. Mag. 3:356. 1818.

19. *Camassia scilloides* (Wild Hyacinth). *a.* Inflorescence and leaves, X¼. *b.* Flower, X1. *c.* Seed, X2½.

Lemotrys hyacinthina Raf. Fl. Tellur. 3:51. 1836.

Camassia fraseri Torr. Pac. R. R. Rep. 2(4):176. 1855.

Scilla fraseri (Torr.) Gray, Man. Bot. 469. 1856.

Quamasia hyacinthina (Raf.) Britt. in Britt. & Brown, Ill. Fl. 1:423. 1896.

Camassia esculenta (Ker) Robinson, Rhodora 10:31. 1908.

Bulbs tunicated; leaves basal, linear, keeled, to 35 cm long, to 1 cm broad; scapes to 60 cm tall, with 0–2 (–3) bracts deciduous after anthesis; lowest bract 1–8 (–11.5) cm below the lowest pedicel of the inflorescence; inflorescence 3–5 cm broad at anthesis; perianth parts white to pale blue to pale lilac, 10–17 mm long, tapering to the base, sometimes short-clawed, glabrous; style (4.5–) 5.5–8.0 (–11.0) mm long; capsule subgloboid, angular, as broad as long. 2n = 30 (unpublished data).

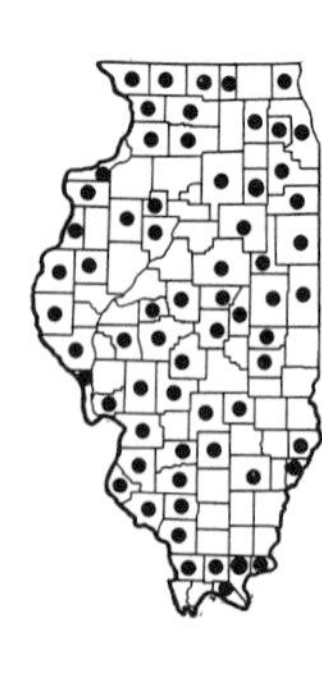

COMMON NAME: Wild Hyacinth.

HABITAT: Moist prairies and woodlands, particularly in calcareous areas.

RANGE: Pennsylvania to Michigan, south to Texas and Georgia.

ILLINOIS DISTRIBUTION: Occasional; throughout the state.

This wild hyacinth flowers from early April until the last of June, often beginning at least a month before *C. angusta.*

Considerable variation exists in flower color, the flowers ranging from the typical pale lilac to the more infrequently encountered pale blue or white.

This species is abundant both in hill prairies as well as along railroad prairies. It also occurs along the edges of woods, particularly in calcareous areas.

2. **Camassia angusta** (Engelm. & Gray) Blankinship, Rept. Mo. Bot. Gard. 18:195. 1907. *Fig. 20.*

Scilla angusta Engelm. & Gray, Boston Jour. Nat. Hist. 5:237. 1845.

Anthericum serotinum Baker, Bot. Jahrb. 15:beibl. 35:9. 1892, non L. (1762).

Bulbs tunicated; leaves basal, linear, keeled, to 35 cm long, to 1 cm broad; scapes to 60 cm tall, with 3–24 bracts persistent

20. *Camassia angusta* (Wild Hyacinth). *a*. Inflorescence and leaves, X¼. *b*. Flower, X⅝. *c*. Capsule, X1. *d*. Seed, X2½.

after anthesis; lowest bract (2–) 5–40 cm below the lowest pedicel of the inflorescence; inflorescence 2–3 (–3.5) cm broad at anthesis; perianth parts deep lavender to pale purple, 7–10 (–13) mm long, tapering to the base, sometimes short-clawed, glabrous; style 2.5–4.5 (–5.5) mm long; capsule longer than broad.

COMMON NAME: Wild Hyacinth.

HABITAT: Prairies and woodlands.

RANGE: Illinois to Kansas, south to Texas and Arkansas.

ILLINOIS DISTRIBUTION: Very rare; known only from Macon County. Steyermark (1961) has pointed out the validity of this forgotten species. The darker flowers, longer capsules, and numerous bracts, together with the later flowering time, distinguish this species from the more common *C. scilloides*.

14. *Erythronium* L. – Dog-tooth Violet

Inflorescence solitary, terminal; flowers perfect; perianth parts 6, free; style 1; capsule loculicidal; bulbous plants with 1–2 basal leaves.

The paired leaves and solitary, terminal flower are the distinguishing generic characters.

KEY TO THE SPECIES OF Erythronium IN ILLINOIS

1. Flowers yellow; stigmas united______________1. *E. americanum*
1. Flowers white; stigmas free_____________________2. *E. albidum*

1. Erythronium americanum Ker, Bot. Mag. pl. 1113. 1808. *Fig. 21.*

Bulb scaly, deep in the ground; leaves 1–2, basal, elliptic to elliptic-lanceolate, acute or subacute, to 15 cm long, to 4 (–5) cm broad, glabrous, green, usually mottled; peduncle 5–15 cm long, glabrous, arching; flower solitary, nodding, yellow, frequently spotted near the base; perianth parts broadly lanceolate, 1.5–3.5 cm long, the inner three bearing a pair of marginal glands near the base; stigmas united; capsule elongate-subgloboid; 2n = 48 (Parks & Hardin, 1963), 48 (Haque, 1951).

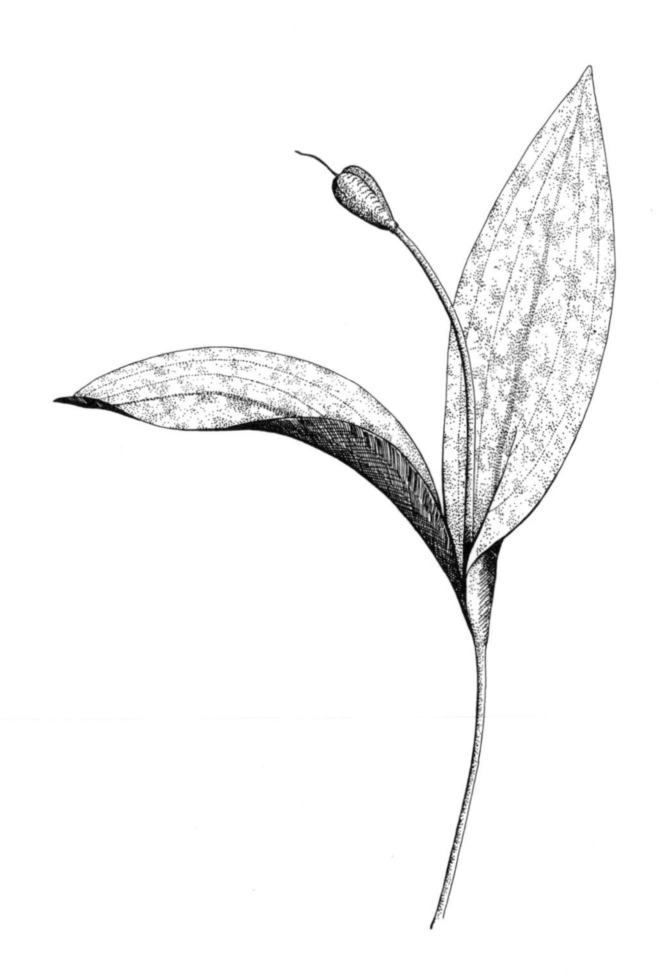

21. *Erythronium americanum* (Yellow Dog-tooth Violet). *a*. Habit, in flower, X¼. *b*. Habit, in fruit, X¼.

COMMON NAME: Yellow Dog-tooth Violet; Yellow Trout Lily.

HABITAT: Moist woodlands, frequently on shaded bluffs.

RANGE: New Brunswick to Ontario, south to Oklahoma and Georgia.

ILLINOIS DISTRIBUTION: Occasional in the northeastern, east-central, and southern counties; apparently absent elsewhere, except Calhoun County.

The yellow dog-tooth violet or trout lily flowers from April to mid-May. The peduncles arise between the paired basal leaves. Single-leaved specimens, which are immature, do not flower.

It is extremely difficult to distinguish this species from the following solely on fruit characters.

2. **Erythronium albidum** Nutt. Gen. N. Am. Pl. 1:223. 1818. *Fig. 22.*

Bulb scaly, deep in the ground; leaves 1–2, basal, elliptic to elliptic-lanceolate, acute or subacute, to 15 cm long, to 3.5 (–4.0) cm broad, glabrous, green, only occasionally mottled; peduncle 5–15 cm long, glabrous, arching; flower solitary, nodding, white or bluish-white, rarely pinkish, more or less yellow at the base; perianth parts broadly lanceolate, 1.5–3.5 cm long, without glands; stigmas free; capsule elongate-subgloboid; 2n = 44 (Cooper, 1939).

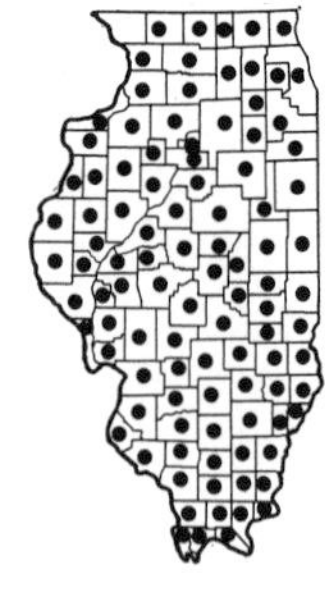

COMMON NAME: White Dog-tooth Violet; White Trout Lily.

HABITAT: Woods and occasionally fields.

RANGE: Ontario to Minnesota, south to Oklahoma and Georgia.

ILLINOIS DISTRIBUTION: Common throughout the state; known from every county except JoDaviess.

This species blooms during April and the first part of May. Its distribution in Illinois covers the entire state. In areas of rocky outcrops, it seems to be less abundant than *E. americanum*.

White dog-tooth violet is found most frequently in woods where it is associated with *Acer saccharum, Fraxinus americana, Quercus rubra, Prunus serotina, Ulmus americana, Clay-*

22. *Erythronium albidum* (White Dog-tooth Violet). *a*. Habit, X¼. *b*. Capsule (immature).

tonia virginica, Dentaria laciniata, Dicentra cucullaria, Galium aparine, Geranium maculatum, Geum canadense, Hydrophyllum virginianum, Isopyrum biternatum, Phlox divaricata, Podophyllum peltatum, Polemonium reptans, Sanicula gregaria, Smilacina racemosa, and *Trillium recurvatum.*

15. *Uvularia* L. – Bellwort

Inflorescence terminal, but appearing more or less axillary at maturity because of the elongation of the branchlets; flowers perfect; perianth parts 6, free; style 1; capsule loculicidal; rhizomatous herbs with cauline leaves.

The cauline leaves, terminal flowers appearing axillary, and perfect flowers characterize this genus.

For a revision of this genus, see Wilbur (1963).

KEY TO THE SPECIES OF Uvularia IN ILLINOIS

1. Leaves perfoliate, puberulent beneath; flowers bright yellow, the perianth parts nearly all over 25 mm long_______1. *U grandiflora*
1. Leaves sessile, not perfoliate, glabrous; flowers stramineous, the perianth parts nearly all less than 25 mm long____2. *U. sessilifolia*

1. Uvularia grandiflora Sm. Exotic Bot. 1:99. 1804. *Fig. 23.*

Rhizomes with long, fleshy, fibrous roots; stems to 50 (–75) cm tall; leaves broadly elliptic to lance-ovate, to 12 cm long, short-acuminate, perfoliate at base, puberulent beneath, the lowest leaves reduced to bladeless sheaths; flowers 1–4, bright yellow, nodding; perianth parts 25–55 mm long, acute; capsule cylindric, obtusely lobed; 2n = 14 (Anderson & Whitaker, 1934).

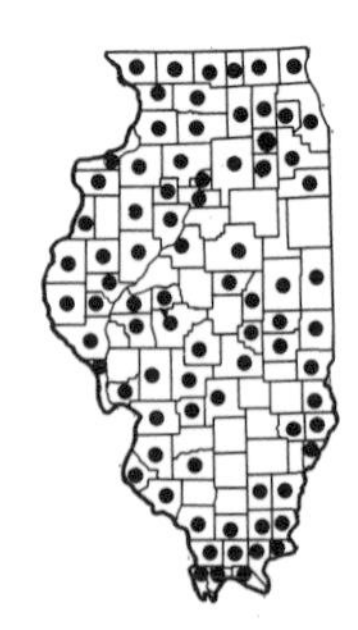

COMMON NAME: Yellow Bellwort.

HABITAT: Rich woodlands.

RANGE: Quebec to North Dakota, south to Oklahoma and Georgia.

ILLINOIS DISTRIBUTION: Rather common; throughout Illinois, but not as yet collected in every county.

This attractive wild flower blooms during April and May. Its perfoliate leaves distinguish it from any other member of the Liliaceae in Illinois. It is invariably associated with *Trillium recurvatum, Asarum reflexum,* and other common early spring wild flowers.

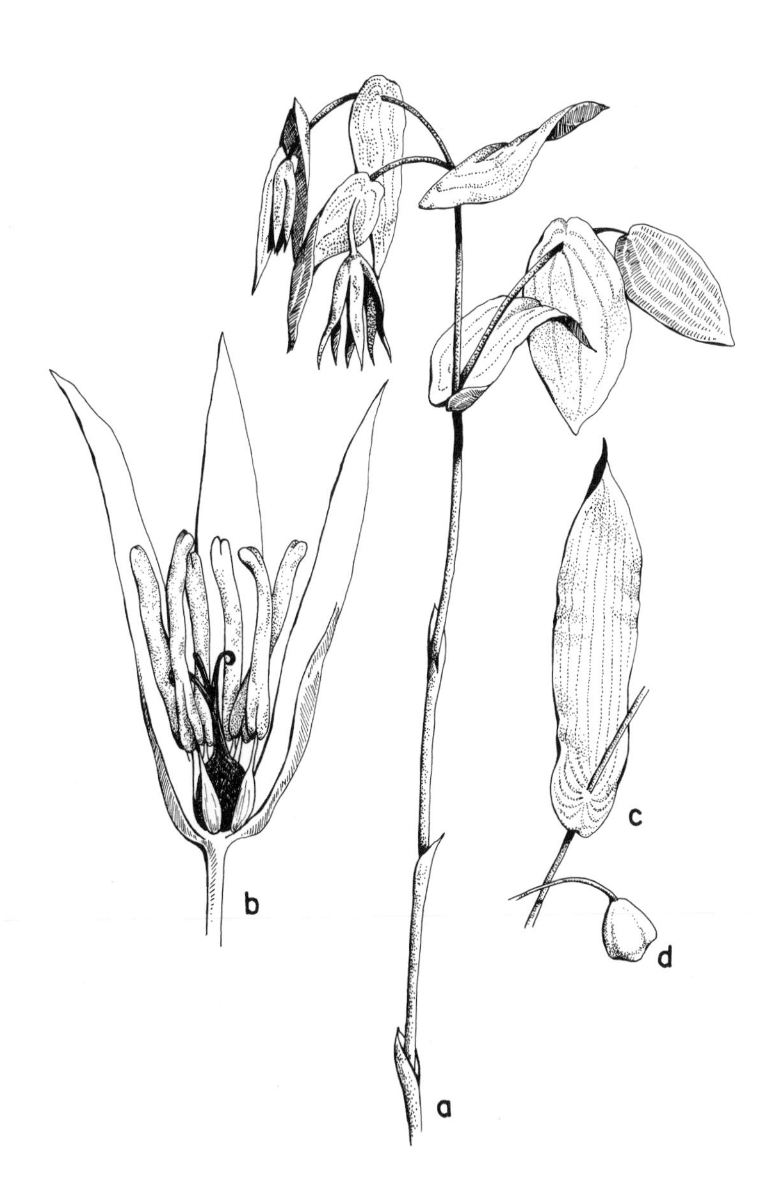

23. *Uvularia grandiflora* (Yellow Bellwort). *a*. Habit, X¼. *b*. Flower (some perianth parts removed), X1⅛. *c*. Leaf, X¼. *d*. Capsule, X¼.

2. **Uvularia sessilifolia** L. Sp. Pl. 305. 1753. *Fig. 24.*

Oakesia sessilifolia (L.) Wats. Proc. Am. Acad. 14:269. 1879.

Oakesiella sessilifolia (L.) Small, Fl. S.E. U.S. 271. 1903.

Rhizome very short, with fleshy, fibrous roots; stems to 30 cm tall, glabrous; leaves broadly lanceolate, to 8 cm long, acute, more or less rounded at the sessile base, glabrous, glaucous beneath, the lowest leaves reduced to bladeless sheaths; flowers 1–3, stramineous, pendent; perianth parts up to 25 mm long, acute; capsule ellipsoid, obtusely lobed, to 2 cm long.

COMMON NAME: Sessile-leaved Bellwort; Small Bellwort.

HABITAT: Rich woodlands.

RANGE: New Brunswick to North Dakota, south to Missouri and Georgia.

ILLINOIS DISTRIBUTION: Not common; confined to the southern one-third of the state.

This delicate spring wild flower blooms during April and the first of May. It is distinguished from *U. grandiflora* by its glabrous and sessile leaves and its stramineous and smaller flowers. It is found in the richer woodlands of southern Illinois. It is one of the associates of the rare orchid, *Isotria verticillata.*

This species was cited by Wilbur (1963) from Cook County. Mr. Floyd Swink of the Morton Arboretum has checked the authenticity of this report and has found that it is based on a collection where a mix-up of labels is fairly certain. The evidence seems to indicate a confusion in localities so that I am excluding *U. sessilifolia* from Cook County.

16. Polygonatum MILL. – Solomon's Seal

Flowers axillary, pendulous, perfect; perianth parts 6, united into a short tube; style 1; berry several-seeded; rhizomatous plants with rather broad cauline leaves.

24. *Uvularia sessilifolia* (Sessile-leaved Bellwort). *a*. Habit, X¼. *b*. Flower, X⅝. *c*. Leaf and capsule, X¼.

25. *Polygonatum pubescens* (Small Solomon's Seal). *a.* Habit, X⅛. *b.* Flower (diagrammatic longitudinal view), X1¾. *c.* Fruiting clusters, X¼.

KEY TO THE SPECIES OF Polygonatum IN ILLINOIS

1. Leaves pilose on the nerves beneath ____________ 1. *P. pubescens*
1. Leaves glabrous beneath.
   2. Leaves more or less clasping or sheathing at the base, the largest ones with over 100 nerves; perianth 17–20 mm long, the lobes 5–7 mm long ____________ 2. *P. commutatum*
   2. Leaves sessile, the largest ones with less than 100 nerves; perianth 10–17 mm long, the lobes 3–4 mm long _____ 3. *P. biflorum*

1. Polygonatum pubescens (Willd.) Pursh, Fl. Am. Sept. 1:234. 1814. *Fig. 25.*

Convallaria pubescens Willd. Hort. Berol. 1:pl. 45. 1805.

Perennial from rather slender superficial rhizomes; stems slender, to 85 (–100) cm tall, glabrous; leaves elliptic-lanceolate, subacute at apex, more or less tapering to the usually sessile base, to 10 (–14) cm long, up to 7 cm broad, pilose on the nerves beneath, usually green on both sides; peduncles slender, glabrous, to 5 (–7) cm long, 1- to 3-flowered; pedicels 2–10 cm long, glabrous; perianth 7–13 mm long, the lobes 3–4 mm long; seeds globoid, smooth, light brown or tan, 1.5–3.0 mm in diameter.

COMMON NAME: Small Solomon's Seal.

HABITAT: Moist, shaded woods.

RANGE: Quebec to Manitoba, south to Iowa, Kentucky, and South Carolina.

ILLINOIS DISTRIBUTION: Restricted to five extreme northern counties.

The pubescent veins on the lower leaf surface distinguish this species of *Polygonatum* from the other species in Illinois. The number of flowers from each axil ranges from one to three; pedicel length is extremely variable, but is primarily a matter of maturity.

2. Polygonatum commutatum (Schult.) A. Dietr. ex Otto & A. Dietr. Allg. Gartenz. 3:223. 1835. *Fig. 26.*

Convallaria commutata Schult. in Roem. & Schult. Syst. Veg. 7(2):1671. 1830.

Polygonatum giganteum A. Dietr. ex Otto & A. Dietr. Allg. Gartenz. 3:222. 1835.

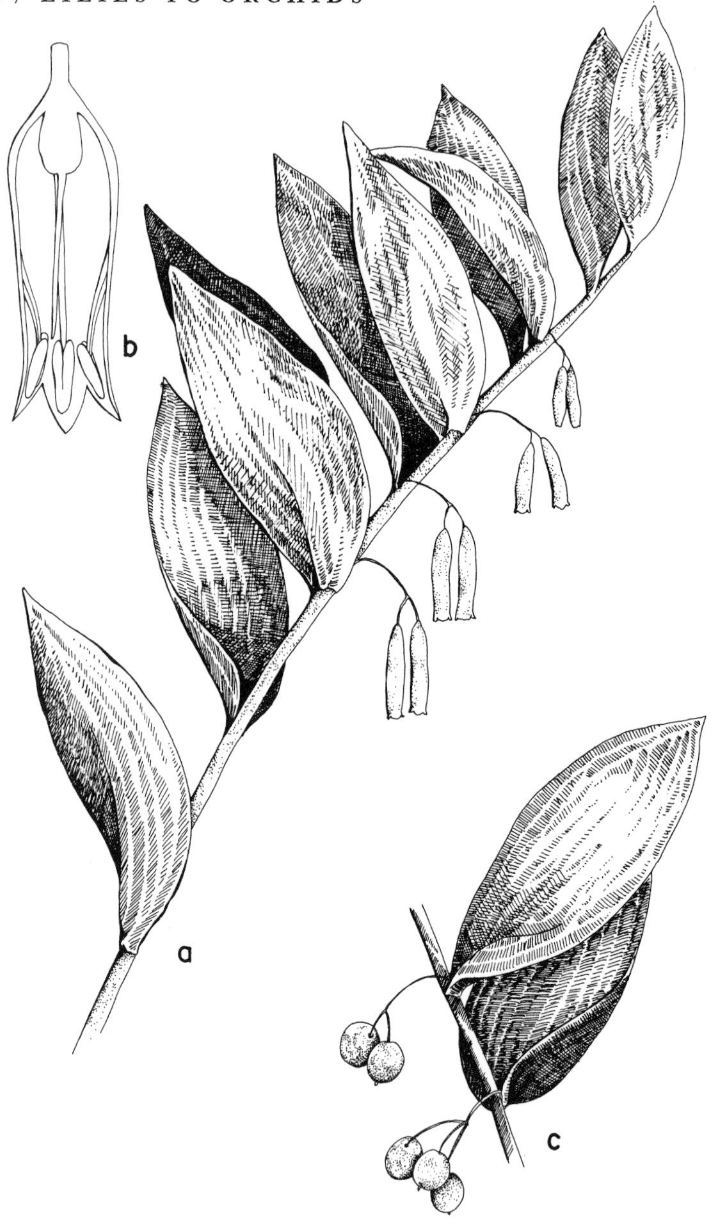

26. *Polygonatum commutatum* (Solomon's Seal). *a.* Habit, X¼. *b.* Flower (diagrammatic longitudinal view), X1⅛. *c.* Fruiting clusters, X¼.

Salomonia commutata (Schult.) Farw. in Rep. Comm. Parks, Detroit 11:53. 1900.

Perennial from stout rhizomes; stems stout, to 2 m tall, glabrous; leaves broadly elliptic to suborbicular, obtuse to subacute at apex, rounded at the more or less clasping or sheathing base, to 15 (–25) cm long, to 10 (–15) cm broad, glabrous and green on both sides, with over 100 nerves on the largest leaves; peduncles stout, glabrous, to 8.5 cm long, 2- to 10-flowered; pedicels to 4 cm long, glabrous; perianth 17–20 mm long, the lobes 5–7 mm long; seeds angular-globoid, smooth, 1.5–3.0 mm in diameter.

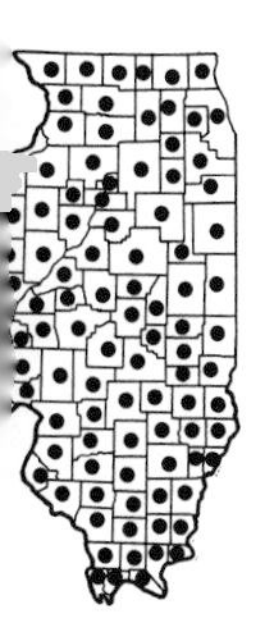

COMMON NAME: Solomon's Seal.

HABITAT: Woods, both dryish and moist.

RANGE: New Hampshire to Manitoba, south to Oklahoma and South Carolina.

ILLINOIS DISTRIBUTION: Common; in every county.

This species is extremely variable. Gigantic species attaining a height of 2 meters and leaves 25 cm long are known. They have been called *P. giganteum.* Small specimens approach *P. biflorum,* but differ in their clasping leaf bases and their longer perianths. *Polygonatum canaliculatum,* sometimes the name used for this species, is a synonym for *P. biflorum.*

This wild flower is familiar to almost everyone who has walked in the woods. It is a common associate of *Geranium maculatum, Phlox divaricata,* and *Polemonium reptans.*

3. **Polygonatum biflorum** (Walt.) Ell. Bot. S. C. & Ga. 1:393. 1817. *Fig. 27.*

Convallaria biflora Walt. Fl. Carol. 122. 1788.

Convallaria canaliculata Willd. Enum. Hort. Berol. 45. 1809.

Polygonatum canaliculatum (Muhl.) Pursh, Fl. Am. Sept. 1:234. 1814.

Perennial from slender rhizomes; stems slender, to 75 (–90) cm tall, glabrous; leaves elliptic-lanceolate to rarely ovate, subacute at apex, more or less rounded or tapering to sessile base, to 8.5 (–12) cm long, up to 5 cm broad, glabrous on both sides, sometimes glaucous beneath, with up to 100 nerves on the larg-

27. *Polygonatum biflorum* (Small Solomon's Seal). *a.* Habit, X¼. *b.* Flower, X1¼. *c.* Fruiting clusters, X¼.

est leaves; peduncles slender, glabrous, to 4 cm long, 1- to 3-flowered; pedicels to 2 cm long, glabrous; perianth 10–17 mm long, the lobes 3–4 mm long; seeds pale brown, 2.5–3.5 mm in diameter.

COMMON NAME: Small Solomon's Seal.

HABITAT: Dry woods; atop sandstone cliffs.

RANGE: Connecticut to Michigan, south to Nebraska, Texas, and Florida.

ILLINOIS DISTRIBUTION: Restricted to the Shawneetown Ridge in the southern tip of the state.

The binomial *Polygonatum biflorum* in the past has been used to include not only the typical *P. biflorum* but also *P. commutatum.* Ownbey (1944) points out that since not a single character in the description of *Convallaria canaliculata* Muhl. corresponds with the characters of the present *P. commutatum,* it is best to consider *C. canaliculata* as a synonym for *P. biflorum. Convallaria canaliculata* is the basionym for *P. canaliculatum* (Muhl.) Pursh.

Specimens of *P. biflorum* resemble small plants of *P. commutatum,* but differ in their sessile rather than clasping leaf bases and their shorter perianths.

Thus far, this species has been found only on sandstone ridge-tops or on the upper slopes of oak-hickory woodlands.

17. *Smilacina* DESF. – False Solomon's Seal

Inflorescence terminal, racemose or paniculate; flowers perfect; perianth parts 6, free; style 1; berry 1- to 2-seeded; rhizomatous plants with rather broad cauline leaves.

The species of *Smilacina* known to occur in Illinois were placed originally by Linnaeus in *Convallaria,* a genus with united perianth parts and a many-seeded berry.

Confusion sometimes exists between this genus and *Polygonatum,* primarily because of the similarity in common names, but *Smilacina* has a terminal inflorescence while that in *Polygonatum* is axillary.

Although Fernald (1950) records *S. trifolia* (L.) Desf. from "n. Ill.," no specimens have been seen to warrant inclusion of this species in the Illinois flora.

KEY TO THE SPECIES OF Smilacina IN ILLINOIS

1. Flowers paniculate; stamens longer than the perianth parts; perianth parts to 3 mm long; pedicels less than 4 mm long; berry red--1. *S. racemosa*
1. Flowers racemose; stamens shorter than or equalling the perianth parts; perianth parts 4–6 mm long; pedicels at least 4 mm long; berry black or greenish-black--------------------2. *S. stellata*

1. Smilacina racemosa (L.) Desf. Ann. Mus. Paris 9:52. 1807. *Fig. 28.*

Convallaria racemosa L. Sp. Pl. 315. 1753.

Vagnera racemosa (L.) Morong, Mem. Torrey Club 5:114. 1894.

Rhizomes stout, knotty; stem to nearly 1 m long, curved, minutely pubescent; leaves 5–20, elliptic, to 15 cm long, to 7.5 cm broad, 2-ranked, short-acuminate at apex, abruptly tapering at base, glabrous above, minutely pubescent beneath; panicle pyramidal, to 15 cm long, the peduncle minutely pubescent, 1–8 cm long; flowers numerous, white, the pedicels less than 4 mm long; perianth parts linear to narrowly lanceolate, 1.5–3.0 mm long; stamens longer than the perianth parts; berry red, 4–7 mm in diameter.

COMMON NAME: False Solomon's Seal; False Spikenard.

HABITAT: Rich, moist woodlands.

RANGE: Nova Scotia to Alaska, south to Arizona and Georgia.

ILLINOIS DISTRIBUTION: Common; in every county.

This familiar wildflower is one of the most common spring-flowering plants in Illinois. It flowers from the last of April nearly to the end of June.

Although small-flowering specimens of *S. racemosa* tend to resemble *S. stellata,* they may be distinguished by their protruding stamens. In a fruiting condition, the red berry readily distinguishes *S. racemosa.*

The smaller var. *cylindrata* Fern., which may not be worthy of recognition, seems to range outside of Illinois.

28. *Smilacina racemosa* (False Solomon's Seal). *a*. Habit, X¼. *b*. Flower, X1½. *c*. Fruiting clusters, X¼.

2. Smilacina stellata (L.) Desf. Ann. Mus. Paris 9:52. 1807. *Fig. 29.*

Convallaria stellata L. Sp. Pl. 316. 1753.

Vagnera stellata (L.) Morong, Mem. Torrey Club 5:114. 1894.

Rhizomes slender; stem to 60 cm tall, more or less erect, minutely pubescent or glabrous; leaves 6–12, lanceolate to oblong-lanceolate, to 15 cm long, to 4 cm broad, 2-ranked, more or less plicate, acute at apex, tapering to the sessile or subclasping base, glabrous above, minutely pubescent beneath; raceme 5- to 12-flowered, to 5 cm long, subsessile; flowers white, the pedicels 4–6 mm long; perianth parts linear, 4–6 mm long; stamens shorter than the perianth parts; berry black or greenish-black, 6–10 mm in diameter.

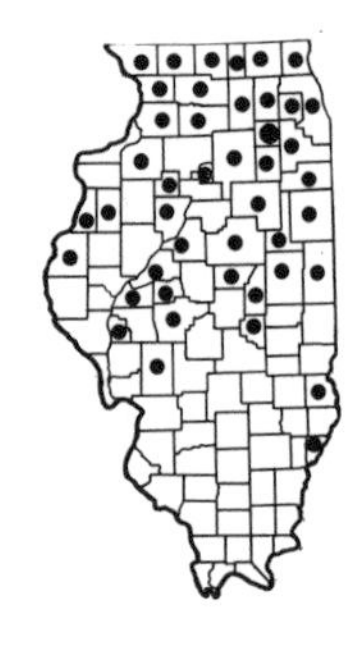

COMMON NAME: Small False Solomon's Seal.

HABITAT: Moist woodlands; prairies.

RANGE: Newfoundland to Alaska, south to California, Missouri, and Virginia.

ILLINOIS DISTRIBUTION: Occasional in the northern three-fifths of Illinois; Crawford and Wabash counties; absent elsewhere. There have been some erroneous reports of this species from the southern tip of Illinois, such as that by Mohlenbrock (1954). Most of the specimens on which these reports are based are immature *S. racemosa.*

Where *Smilacina stellata* occurs in prairies, it is associated with *Comandra richardsiana, Coreopsis palmata, Fragaria virginiana, Monarda fistulosa, Panicum scribnerianum, P. virgatum, Phlox pilosa, Poa pratensis, Tradescantia ohiensis,* and *Zizia aurea.*

This species, of smaller stature than *S. racemosa,* nonetheless has larger flowers which open in early May and continue into June.

18. *Convallaria* L. – Lily-of-the-Valley

Inflorescence terminal, racemose, secund; flowers perfect; perianth parts 6, united into a short tube; style 1; berry many-seeded; rhizomatous plants with 2–3 cauline leaves and fragrant flowers.

29. *Smilacina stellata* (Small False Solomon's Seal). *a.* Habit, X¼. *b.* Flower, X2½. *c.* Inflorescence and upper leaves, X¼.

Only the following species is known from Illinois.

1. **Convallaria majalis** L. Sp. Pl. 314. 1753. *Fig. 30.*

Rhizomes slender, extensively creeping; leaves 2–3, cauline, elliptic, short-acuminate, to 20 cm long, to 7 cm broad, glabrous; scape to 20 cm tall; raceme secund, 6- to 12-flavored, with small, lanceolate bracts; flowers white, fragrant, the pedicels 10–20 mm long, pendulous; perianth 6–10 mm long, campanulate, the shallow lobes broad, obtuse; stamens much shorter than the perianth; capsule globoid, red, 8–10 mm in diameter; seeds globoid; n = 19 (Sorsa, 1962; Matsuura and Sutô, 1935).

COMMON NAME: Lily-of-the-Valley.

HABITAT: Waste ground; in a tamarack bog in one Illinois location.

RANGE: Native of Europe.

ILLINOIS DISTRIBUTION: Not generally escaped; only four Illinois collections seen.

This common garden plant flowers during May and June.

The establishment of a colony on a dry "hump" in the Wauconda tamarack bog is an interesting attribute of its ability to become a naturalized plant.

19. *Asparagus* L. – **Asparagus**

Flowers axillary, unisexual; perianth parts 6, united into a short tube; style 1; berry few-seeded; rhizomatous, dioecious plants with minute, scale-like leaves and short, linear branchlets.

The short, linear branchlets, each subtended by a scale leaf, functionally are the leaves. They are incapable of further lengthening, and frequently become flattened.

Only the following species occurs in Illinois.

1. **Asparagus officinalis** L. Sp. Pl. 313. 1753. *Fig. 31.*

Rhizome rather stout, much branched; stem branched, to nearly 2 m tall, glabrous; leaves reduced to scales, each with a short-linear branchlet in the axil, the branchlet to 15 mm long; flowers solitary (rarely paired), axillary, the pedicels to 10 mm long, jointed at the middle; perianth campanulate, 3–5 mm long, greenish-white, the short lobes subacute to acute; berry red, globoid, 6–8 mm in diameter; 2n = 20 (unpublished data).

30. *Convallaria majalis* (Lily-of-the-Valley). *a.* Habit, X¼. *b.* Flower, X1¼.

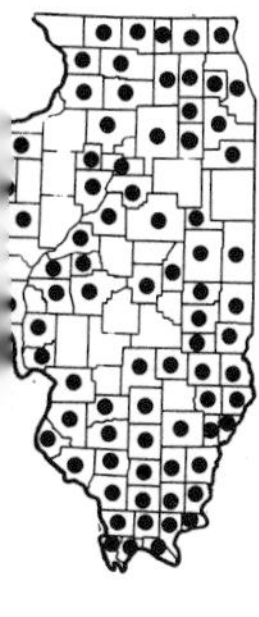

COMMON NAME: Garden Asparagus; Asparagus.
HABITAT: Waste ground.
RANGE: Native to Europe; commonly escaped from cultivation throughout North America.
ILLINOIS DISTRIBUTION: Common; probably in every county, but thus far not collected in all counties. The young stalks are highly sought for a delicate food.

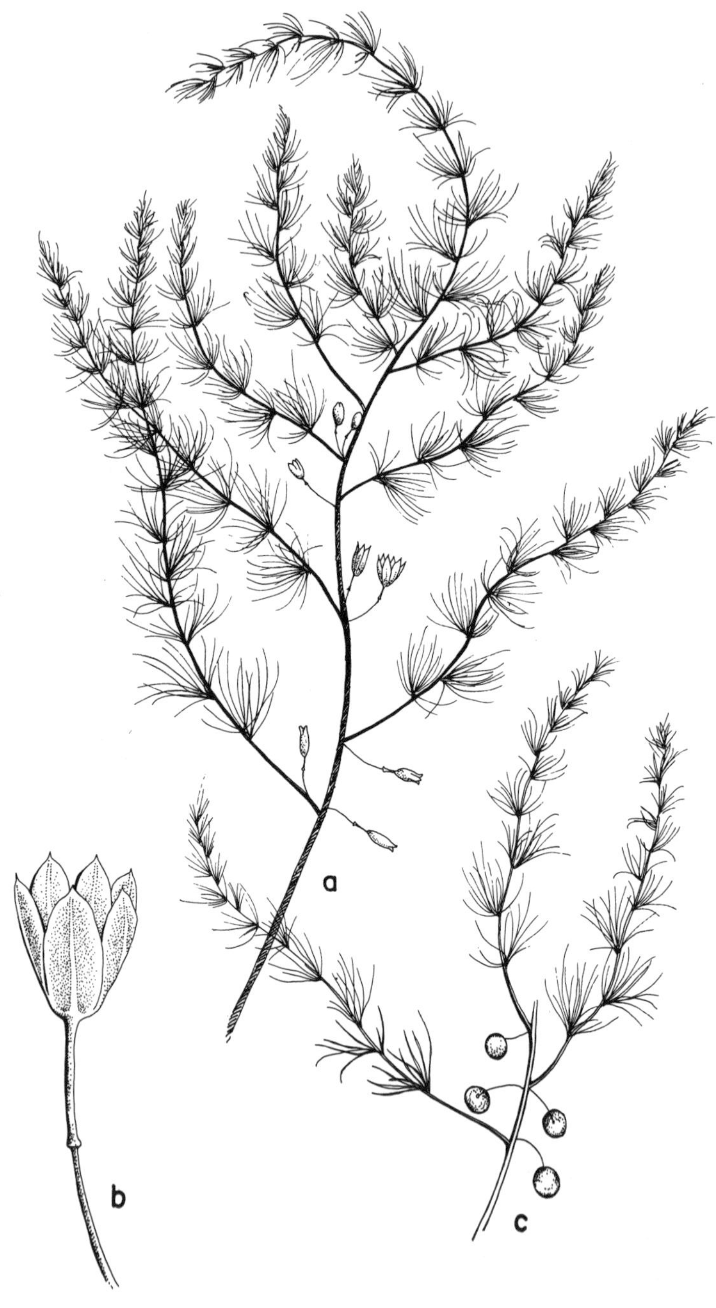

31. *Asparagus officinalis* (Asparagus). *a.* Habit, X½. *b.* Flower, X3. *c.* Fruiting branch, X¼.

20. *Maianthemum* WEBER – False Lily-of-the-Valley

Inflorescence terminal, racemose; flowers perfect; perianth parts 4, free; stamens 4; style 1, bilobed; berry 1- to 2-seeded; rhizomatous plants with 1–3 broad cauline leaves.

Only the following species occurs in Illinois.

1. **Maianthemum canadense** Desf. Ann. Mus. Paris 9:54. 1807. *Fig. 32.*

Rhizome very slender, extensively creeping; stem to 20 cm tall, somewhat flexuous; leaves 1–3, to 10 cm long, to 5 cm broad, oblong-ovate to ovate, obtuse to subacute at the apex, tapering to the sessile or short-petiolate base, except for the somewhat larger lowest, cordate leaf, glabrous or pubescent beneath; raceme erect, to 5 cm long, short-pedunculate; flowers white, fragrant, the pedicels very short; perianth parts linear-lanceolate, subacute, 2.0–2.5 mm long; berry light red, 3–4 mm in diameter, 1- to 2-seeded.

COMMON NAME: False Lily-of-the-Valley.

The 2- to 4-parted flowers are unique among the Liliaceae. The fragrant flowers are reminiscent of the true lily-of-the-valley.

Two varieties occur in Illinois, separated below. There is some evidence to support the view that these two varieties should be considered separate species.

a. Leaves glabrous beneath; transverse veins of leaf conspicuous in transmitted light______________1a. *M. canadense* var. *canadense*
a. Leaves pubescent beneath; transverse veins of leaf obscure in transmitted light____________________1b. *M. canadense* var. *interius*

1a. **Maianthemum canadense** Desf. var. **canadense**

HABITAT: Moist woodlands.

RANGE: Labrador to Minnesota, south to Pennsylvania; in the mountains of North Carolina, Tennessee, and Georgia.

ILLINOIS DISTRIBUTION: Rare; only known from a single collection from Cook County.

32. *Maianthemum canadense* (False Lily-of-the-Valley).—var. *canadense*. *a*. Habit, X¼. *b*. Flower, X3¾. *c*. Fruiting branch, X¼.—var. *interius*. *d*. Habit, X¼. *e*. Fruiting branch, X¼.

1b. Maianthemum canadense var. **interius** Fern. Rhodora 16:211. 1914.

HABITAT: Moist woodlands.
RANGE: Ontario to British Columbia, south South Dakota, Iowa, and Massachusetts.
ILLINOIS DISTRIBUTION: Occasional; restricted to nine counties in the northern one-fourth of the state.
The pubescent variety is the more common one in Illinois. It flowers during May and June. Both varieties occur in densely shaded woods.

21. *Allium* L. – Onion

Inflorescence umbellate; flowers perfect; perianth parts 6, free or attached at their bases; style 1; ovary 3-celled, with 1–2 ovules per cell; capsule loculicidal; bulbous plants with a pungent odor and generally with basal leaves.

Several species of *Allium* are grown in cultivation, and a few of these escape into wasteland.

In some species, the flowers may be replaced by bulblets. For accurate identification of the species of *Allium*, bulbs, leaves, flowers, bulblets, and capsules need to be present.

KEY TO THE TAXA OF Allium IN ILLINOIS

1. Leaves absent at flowering time, some or all of them, when present, over 2.5 cm broad ---------------------------- 1. *A. tricoccum*
1. Leaves present at flowering time, up to 1.5 cm broad.
  2. Leaves flat or channelled, not hollow.
    3. Leaves extending about half-way up the stem; flowers whitish, greenish, or deep purple.
      4. Umbel producing bulblets; flowers whitish or greenish -- 2. *A. sativum*
      4. Umbel not producing bulblets; flowers deep purple ---------------------- 3. *A. ampeloprasum* var. *atroviolaceum*
    3. Leaves basal or nearly so; some of the flowers usually pinkish.
      5. Umbel producing bulblets ------------ 4. *A. canadense*
      5. Umbel not producing bulblets.
        6. Stems solid -------------------------- 5. *A. porrum*
        6. Stems hollow.

7. Ovary and capsule crested near apex (Fig. 38); outer bulb scales membranous; perianth parts shorter than the stamens at maturity.
    8. Umbel nodding; leaves soft______6. *A. cernuum*
    8. Umbel erect; leaves stiff________7. *A. stellatum*
7. Ovary and capsule without crests; outer bulb scales fibrous; perianth parts usually longer than the stamens at maturity________________8. *A. mutabile*

2. Leaves terete, hollow.
    9. Leaves extending nearly to middle of stem; umbel bulblet-bearing_________________________________9. *A. vineale*
    9. Leaves basal or nearly so; umbel sometimes bulblet-bearing only in number 11.
        10. Flowers pink or purplish, the pedicels shorter than or barely equalling the flowers; umbel not bulblet-bearing; stem not strongly inflated below the middle____10. *A. schoenoprasum*
        10. Flowers white or greenish, the pedicels much longer than the flowers; umbel sometimes bulblet-bearing; stem strongly inflated below the middle________11. *A. cepa*

1. Allium tricoccum Ait. Hort. Kew. 1:428. 1789. *Fig. 33.*

Bulbs ovoid, to 6 cm long, with fibrous outer scales; leaves elliptic-lanceolate, to 20 cm long, to 6 cm broad, flat, appearing in spring, usually absent by flowering time, the petiole to 8 cm long; scape to 50 cm tall, erect; umbel many-flowered, subtended by two ovate, deciduous bracts; flowers white, the pedicels 1–2 cm long, spreading or ascending; perianth parts narrowly ovate, 5–7 mm long, obtuse; stamens equalling the perianth parts; capsule deeply 3-lobed.

Two varieties occur in Illinois, distinguished by the following key:

1. Petioles and leaf sheaths reddish; blades mostly 2.6–6.0 cm broad, elliptic________________________1a. *A. tricoccum* var. *tricoccum*
1. Petioles and leaf sheaths greenish or whitish; blades mostly 0.8–2.0 cm broad, lanceolate____________1b. *A. tricoccum* var. *burdickii*

1a. Allium tricoccum Ait. var. **tricoccum**

All parts of plant larger in all respects than var. *burdickii;* petioles and leaf sheaths reddish; blades elliptic.

33. *Allium tricoccum* (Wild Leek). *a.* Habit, in fruit, X¼. *b.* Leaves, X¼. *c.* Inflorescence, X¼. *d.* Flower, X1¼. *e.* Capsule, X1¼. *f.* Habit (shaded), X1⁄48.

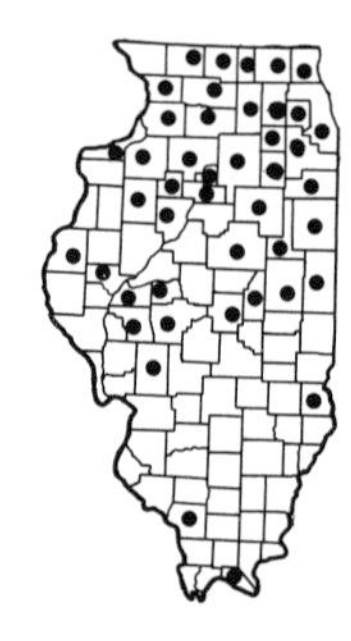

COMMON NAME: Wild Leek; Ramp.
HABITAT: Moist, rich woodlands.
RANGE: New Brunswick to Minnesota, south to Iowa, Tennessee, and Georgia.
ILLINOIS DISTRIBUTION: Occasional in the northern one-half of Illinois; known only from Crawford, Massac, and Jackson counties in the southern one-half.

It was not until 1948 that Hanes and Ownbey reported differences in populations of *Allium tricoccum* in Michigan. Later Hanes (1953) reported the same differences from Wisconsin and Illinois, as well. Typical *A. tricoccum* of upland woods of beech and maple possesses reddish petioles and leaf sheaths, is generally larger in all respects, and produces leaves about one week earlier than var. *burdickii*. On the other hand, var. *tricoccum* comes into flower about one week later than var. *burdickii*.

This interesting plant flowers during June and July, just following the shriveling of the broad leaves. Its rich, mesic upland woods habitat is unique among *Allium* species in Illinois. The huge bulb has an extremely strong flavor.

1b. Allium tricoccum Ait. var. **burdickii** Hanes, Rhodora 55:243. 1953.

All parts of plant smaller in all respects than var. *tricoccum;* petioles and leaf sheaths greenish or whitish; blades lanceolate.

COMMON NAME: Wild Leek; Ramp.
HABITAT: Low, moist woods.
RANGE: Michigan, Wisconsin, Illinois, Missouri.
ILLINOIS DISTRIBUTION: In several counties in the northern one-half of the state.

Further study of var. *burdickii* should reveal a much broader geographical distribution.

2. Allium sativum L. Sp. Pl. 296. 1753. *Fig. 34.*

Bulbs ovoid, several, to 4 cm in diameter, with membranous outer scales; leaves extending nearly half-way up the stem, linear, to 20 cm long, to 1.5 cm broad, flat, keeled, scabrous on the margins and keel, sessile; stem to nearly 1 m tall, erect; umbel relatively few-flowered, subtended by a single bract with a beak to 10 cm long; flowers white or greenish, often replaced by bulblets, the pedicels much longer than the flowers; perianth parts lanceolate, 3–5 mm long; stamens shorter than the perianth parts; capsule shallowly 3-lobed.

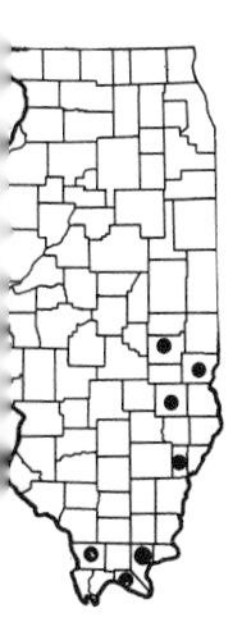

COMMON NAME: Garlic.

HABITAT: Waste areas.

RANGE: Native to Europe and western Asia; infrequently escaped from cultivation.

ILLINOIS DISTRIBUTION: Occasionally escaped throughout the state.

Very often most of the flowers are replaced by ovoid bulblets. Flowers, when present, are not pinkish, as is the case with most species of *Allium* in Illinois.

3. Allium ampeloprasum L. var. **atroviolaceum** (Boiss.) Regel, Acta Horti Petrop. 3(2):54. 1875. *Fig. 35.*

Allium atroviolaceum Boiss. Diagn. Pl. Orient. Nov. 1(7): 112. 1846.

Bulbs ovoid, 2–4 cm long, with papery scales, with numerous bulblets borne from base; leaves flat or sometimes plicate, thick, to 3 cm broad, scabrous on the margins and the keel, hooded at the acute or acuminate apex, long-sheathing; scape to nearly 2 m tall (shorter in the Illinois collection), stout, scaberulous; umbel globose, many-flowered, to 8 cm in diameter, subtended by a globose, scarious, long-beaked bract; flowers deep purple, the pedicels as long as or longer than the flowers; perianth parts oblong-lanceolate, 5–7 mm long, smooth on the back; stamens exserted beyond the perianth parts, the filaments bearing a central anther and two lateral awns terminally; capsule rarely produced.

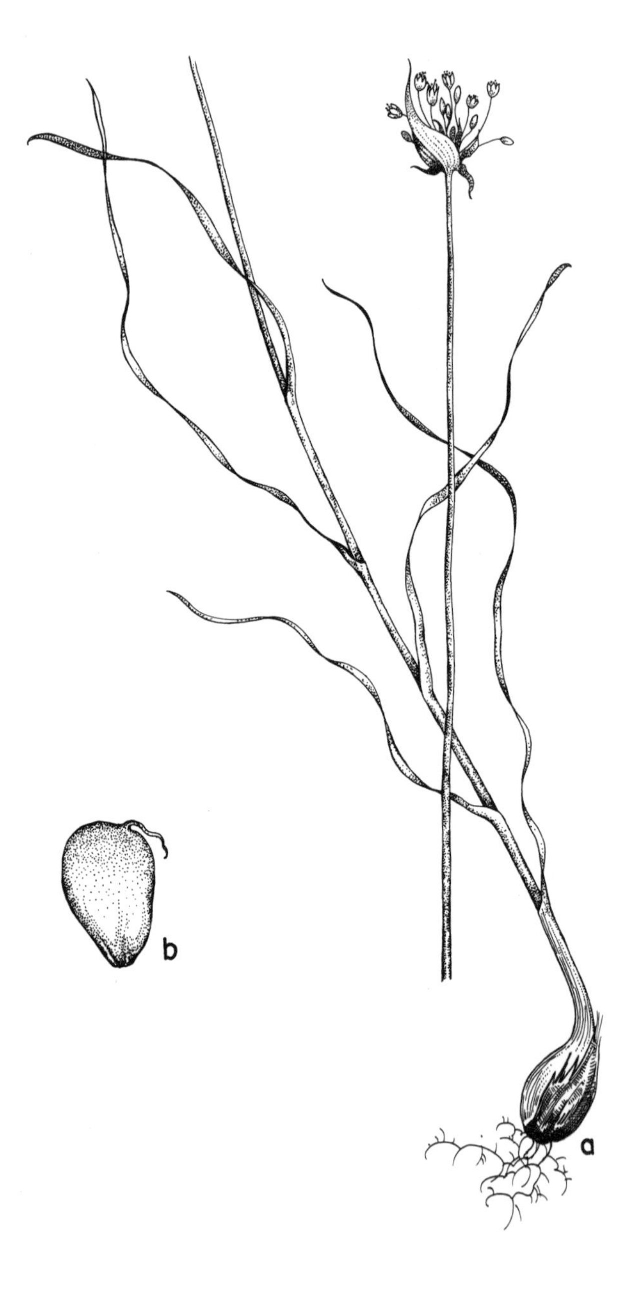

34. *Allium sativum* (Garlic). *a.* Leaves and inflorescence, X¼. *b.* Bulblet, X1½.

35. *Allium ampeloprasum* var. *atroviolaceum* (Wild Leek). *a*. Habit, X⅛. *b*. Flower, X2½. *c*. Capsule, X1½.

COMMON NAME: Wild Leek.
HABITAT: Escaped along roadside.
RANGE: Native of southeast Europe and Asia.
ILLINOIS DISTRIBUTION: Known only from Union (along road in Pine Hills) and Pope (Lusk Creek) counties.
Typical *A. ampeloprasum* of the Old World differs from var. *atroviolaceum* by its smaller umbels and its generally greenish or white perianth parts.

4. **Allium canadense** L. Sp. Pl. 294. 1753. *Fig. 36.*

Bulbs ovoid, to 3 cm long, with fibrous outer scales; leaves basal or nearly so, linear, to 20 cm long, to 5 mm broad, flat, scarcely keeled, smooth, sessile; scape to 50 (–60) cm tall, erect; umbel relatively few-flowered, subtended by (2) –3 ovate, acuminate bracts; flowers pinkish, often entirely replaced by bulblets, the pedicels to 30 mm long, much longer than the flowers; perianth parts lanceolate, acute, to 10 mm long, usually much shorter; stamens shorter than to as long as the perianth parts; capsule subgloboid, shallowly 3-lobed; 2n = 16 (unpublished data).

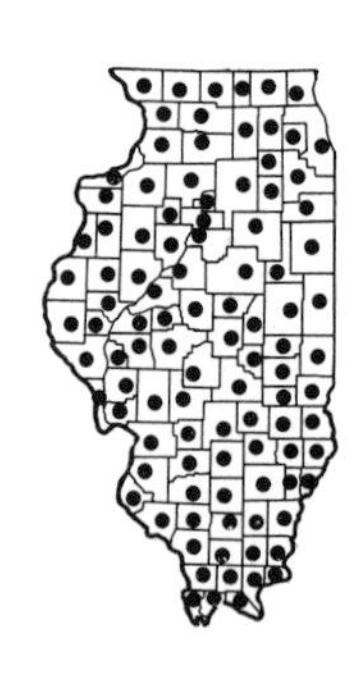

COMMON NAME: Wild Garlic.
HABITAT: Dry woodlands, prairies, waste ground.
RANGE: New Brunswick to South Dakota, south to Texas and Florida.
ILLINOIS DISTRIBUTION: Common; in every county.
This abundant Illinois species frequently has all the flowers replaced by bulblets. When flowers are formed, they appear in May and June. This species is found in a number of basically dry habitats. This is the most common *Allium* with flat leaves in Illinois.

Although this species may become a pest, it is not as obnoxious as *A. vineale*.

5. **Allium porrum** L. Sp. Pl. 295. 1753. *Fig. 37.*

Bulbs ovoid, with papery outer scales; leaves basal or nearly so, linear to elongate-lanceolate, to 25 cm long, to 2 cm broad, flat, keeled, smooth, sessile; scape usually at least 1 m tall, solid, erect; umbel many-flowered, large, subtended by a single,

36. *Allium canadense* (Wild Garlic). *a*. Leaves and inflorescence, X¼. *b*. Flower, X1¼. *c*. Bulblet, X1½. *d*. Inflorescence with flowers and bulblets, X¼.

37. *Allium porrum* (Leek). *a.* Habit (shaded), X1/28. *b.* Inflorescence, X1/4. *c.* Capsule, X1 1/4.

beaked bract; flowers pinkish or whitish, not producing bulblets, the pedicels usually a little longer than the flowers; perianth parts broadly lanceolate, subacute, to 10 mm long; stamens longer than the perianth parts; capsule 3-lobed, rarely produced; $2n = 32$ (Levan, 1931).

COMMON NAME: Leek.
HABITAT: Waste ground.
RANGE: Cultivated plant with unknown origin.
ILLINOIS DISTRIBUTION: Rare as an escaped plant; one collection is known from southern Illinois.
The solid stem distinguishes this species from other large umbellate species in Illinois. The flowers appear from late May to July.

6. **Allium cernuum** Roth ex Roem. Archiv. Bot. 1(3):40. 1789. *Fig. 38.*

Bulbs usually several, narrowly ovoid, with membranous outer scales; leaves basal or nearly so, linear, to 30 cm long, to 5 (–7) mm broad, flat, smooth, sessile; scape to 60 cm tall, hollow, arched near the summit; umbel several-flowered, subtended by 2 bracts; flowers usually pinkish, not producing bulblets, the descending pedicels to 25 mm long; perianth parts narrowly ovate to ovate, subacute, 4–6 mm long; stamens much longer than the perianth parts; capsule shallowly 3-lobed, with each lobe bearing 2 triangular crests; $2n = 14$ (Levan, 1935).

COMMON NAME: Nodding Onion.
HABITAT: Wooded banks.
RANGE: New York to British Columbia, south to Arizona, Texas, and Georgia.
ILLINOIS DISTRIBUTION: Not common; restricted to 11 counties in northeastern Illinois.
The nodding onion is one of the last species of *Allium* to begin flowering in Illinois, usually opening its flowers from July to September. It is related closely to *A. stellatum*, another late-flowering species which differs in its erect umbels and its stiffer leaves.

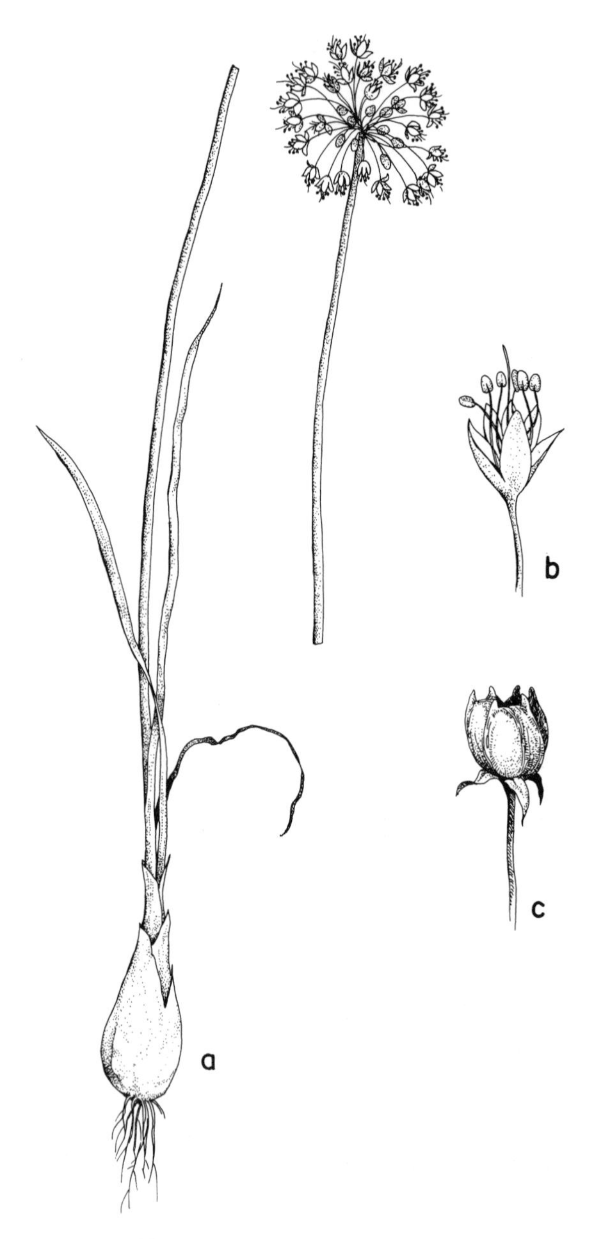

38. *Allium cernuum* (Nodding Onion). *a*. Inflorescence and leaves, X¼. *b*. Flower, X1¼. *c*. Capsule, X1¼.

7. **Allium stellatum** Ker, Bot. Mag. 38:pl.1576. 1813. *Fig.* 39.

Bulbs 1–2, narrowly ovoid, with membranous outer scales; leaves basal or nearly so, linear, to 25 cm long, to 5 mm broad, flat, channelled, smooth, firm, sessile; scape to 50 cm tall, hollow, erect; umbel several-flowered, subtended by 2 lanceolate-ovate, acuminate bracts; flowers pink, not producing bulblets, the ascending pedicels to 20 cm long; perianth parts narrowly ovate, 5–7 mm long; stamens as long as or longer than the perianth parts; capsule shallowly 3-lobed, with each lobe bearing 2 triangular crests.

COMMON NAME: Cliff Onion.

HABITAT: Hill prairies in calcareous regions.

RANGE: Ontario to Saskatchewan, south to Texas and Illinois.

ILLINOIS DISTRIBUTION: Limited to one extreme northern county and four southwestern counties bordering the Mississippi River; abundant where found in the last four counties.

The 6-crested ovary and capsule of this and the preceding species are unique among the Illinois species of *Allium*. *Allium stellatum* differs from *A. cernuum* in its erect umbel and stiff leaves.

8. **Allium mutabile** Michx. Fl. Bor. Am. 1:195. 1803. *Fig.* 40.

Bulbs ovoid, to 3 cm long, with fibrous outer scales; leaves basal or nearly so, linear, to 20 cm long, to 5 mm broad, flat, channelled, smooth, sessile; scape to 50 cm tall, erect, hollow; umbel several-flowered, subtended by 2–3 ovate, short-acuminate bracts; flowers pink, not producing bulblets, the pedicels to 10 mm long in flower; perianth parts broadly lanceolate, acute to short-acuminate, 5–8 mm long; stamens generally longer than the perianth parts; capsule subgloboid, shallowly 3-lobed, without crests.

39. *Allium stellatum* (Cliff Onion). *a.* Inflorescence and leaves, X¼. *b.* Flower, X1¾. *c.* Capsule, X1¼.

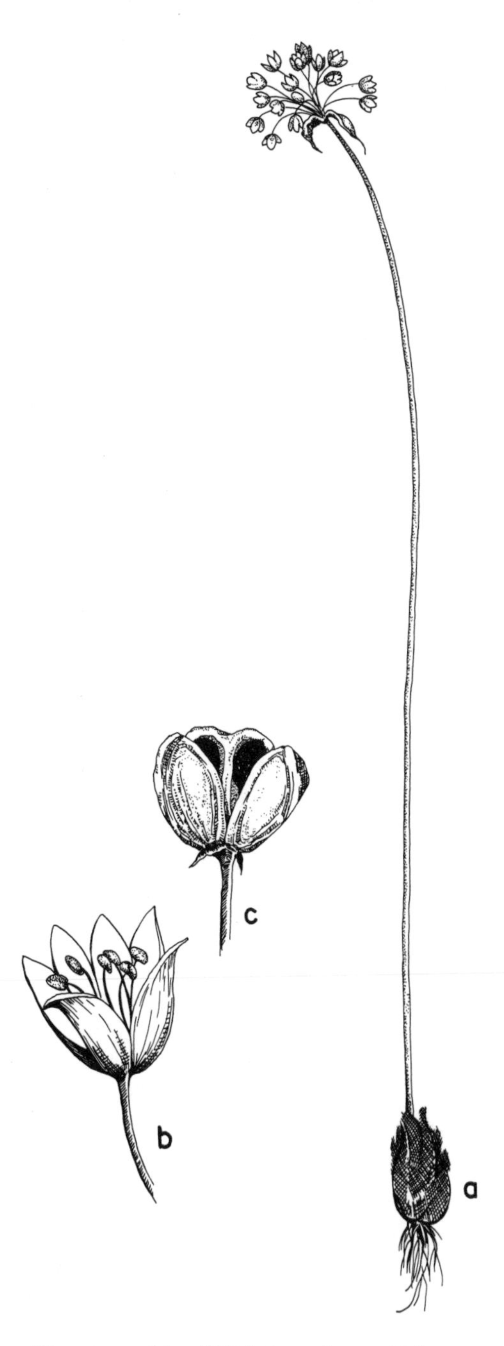

40. *Allium mutabile* (Wild Onion). *a*. Habit, X¼. *b*. Flower, X1¾. *c*. Capsule, X1¼.

COMMON NAME: Wild Onion.

HABITAT: Rather dry situations.

RANGE: North Carolina to Illinois and Nebraska, south to Texas and Florida.

ILLINOIS DISTRIBUTION: Rare; known from only three counties (DuPage, Johnson, and Will) where it was first collected in Illinois from DuPage County by Umbach.

The Johnson County station is based upon a single collection from sandstone bluffs near Trigg Lookout Tower by *J. W. Voigt* (*1109*) in 1952. At the Will County station where it is more abundant, *Allium mutabile* is associated with *Apocynum sibiricum, Arenaria patula, Croton capitatus, Draba reptans, Isanthus brachiatus, Opuntia humifusa, Petalostemum purpureum, Sanguisorba minor, Satureja arkansana, Scutellaria parvula,* and *Verbena simplex.*

Ownbey and Aase (1955) contend that *A. mutabile* is a completely flowering stage of *A. canadense.* This seems to be plausible when one compares the other characters of *A. mutabile* with *A. canadense.*

The colony which occurs in Johnson County has been observed to flower in June. The report by Kibbe (1952) from Hancock County could not be substantiated. The fibrous coat which covers the bulb is very distinguishing.

9. **Allium vineale** L. Sp. Pl. 299. 1753. *Fig. 41.*

Bulbs several, ovoid, to 2 cm long, with membranous outer scales; leaves extending nearly half-way up the stem, to 20 cm long, terete, smooth, sessile; stem sometimes up to 1 m tall, erect; umbel several-flowered, subtended by a single, beaked bract; flowers pinkish, greenish, or whitish, often replaced by bulblets, the pedicels to 20 mm long; perianth parts lanceolate, acute, 3–6 mm long; stamens about as long as the perianth parts, the inner filaments terminated by two hair-like appendages; capsule shallowly 3-lobed.

41. *Allium vineale* (Field Garlic). *a.* Inflorescence and leaves, X¼. *b.* Flower, X1¾. *c.* Bulblet, X1½.

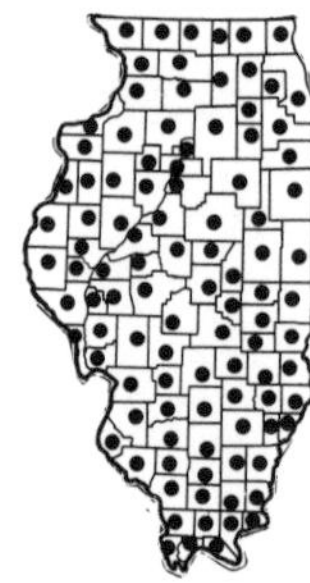

COMMON NAME: Field Garlic.
HABITAT: Waste ground.
RANGE: Native of Europe.
ILLINOIS DISTRIBUTION: Common; in every county.
This obnoxious weed penetrates lawns and fields throughout the state. Variation occurs in the production of bulblets in the inflorescence, with some specimens bearing very few bulblets and others being composed only of bulblets.

Allium vineale, *A. ampeloprasum* var. *atroviolaceum*, and *A. sativum* are the only species of *Allium* in Illinois with leaves borne half-way up the stem. Flowers are produced from late May into August.

10. Allium schoenoprasum L. var. **schoenoprasum** *Fig. 42.*

Allium schoenoprasum L. Sp. Pl. 301. 1753.
Bulbs several, narrowly ovoid, to 2 cm long, 1–2 mm broad, with membranous outer scales; leaves basal or nearly so, to 25 cm long, terete, smooth, sessile; scape to 20 cm tall, erect; umbel many-flowered, subtended by 2 ovate bracts; flowers pink or purple, not producing bulblets, the pedicels 10–15 mm long; perianth parts ovate or lanceolate, acuminate, 10–15 mm long; stamens shorter than the perianth parts; capsule ovoid, shallowly 3-lobed; 2n = 16 (Sokolovskaya, 1963); n = 8 (Matsuura & Sutô, 1935).

COMMON NAME: Chives.
HABITAT: Waste ground.
RANGE: Native of Europe; rarely escaped from cultivation.
ILLINOIS DISTRIBUTION: Rare; collected in the escaped condition only from Jackson County (Carbondale, *R. H. Mohlenbrock 14211*). Variety *sibiricum* (L.) Hartm., which is native to northern North America, is generally larger in all its parts.

Chives flowers from June to August. It is used occasionally as a savory herb.

42. *Allium schoenoprasum* var. *schoenoprasum* (Chives). *a.* Habit, X¼. *b.* Flower, X1¼. *c.* Capsule, X1¼. *d.* Seed, X2½.

11. Allium cepa L. Sp. Pl. 300. 1753. *Fig. 43.*

Bulbs flattened-globoid; leaves basal or nearly so, terete, up to 15 mm in diameter, smooth, sessile; scape up to 1 m tall, broadly inflated below the middle, terete, erect; umbel many-flowered, large, subtended by usually a single bract; flowers white or greenish, sometimes replaced by bulblets, the pedicels much longer than the flowers; perianth parts lanceolate, acute, 3–6 mm long; stamens longer than the perianth parts; 2n = 16 (Battaglia, 1963.)

COMMON NAME: Onion.
HABITAT: Waste ground.
RANGE: Native of southwestern Asia.
ILLINOIS DISTRIBUTION: Rarely escaped from cultivation; our only escaped specimen comes from Union County. The onion is cultivated in Illinois, but rarely escapes. It flowers from late May until August. It is distinguished by its white or greenish flowers, its terete leaves, and its strongly inflated stem.

22. *Nothoscordum* KUNTH – False Garlic

Inflorescence umbellate; flowers perfect; perianth parts 6, attached near their bases; style 1; ovary 3-celled, with 6–10 ovules per cell; capsule loculicidal; bulbous plants without a pungent odor and with basal leaves.

Only the following species occurs in Illinois.

1. Nothoscordum bivalve (L.) Britt. in Britt & Brown, Ill. Fl. 1:415. 1896. *Fig. 44.*

Ornithogalum bivalve L. Sp. Pl. 306. 1753.
Allium striatum Jacq. Coll. Suppl. 51. 1796.

Bulbs ovoid, to 2 cm long, with fibrous outer scales; leaves basal, linear, to 20 cm long, to 3 mm broad, flat, smooth, sessile; scape erect, to 30 cm tall; umbel 5- to 12-flowered, subtended by 2 broadly lanceolate, subacute bracts; flowers white to greenish, not producing bulblets, the pedicels ascending, 20–40 mm long; perianth parts oblong, subacute, 10–14 mm long; stamens about as long as the perianth parts; capsule subgloboid, 4–5 mm long; seeds black, ellipsoid, 3–4 mm long.

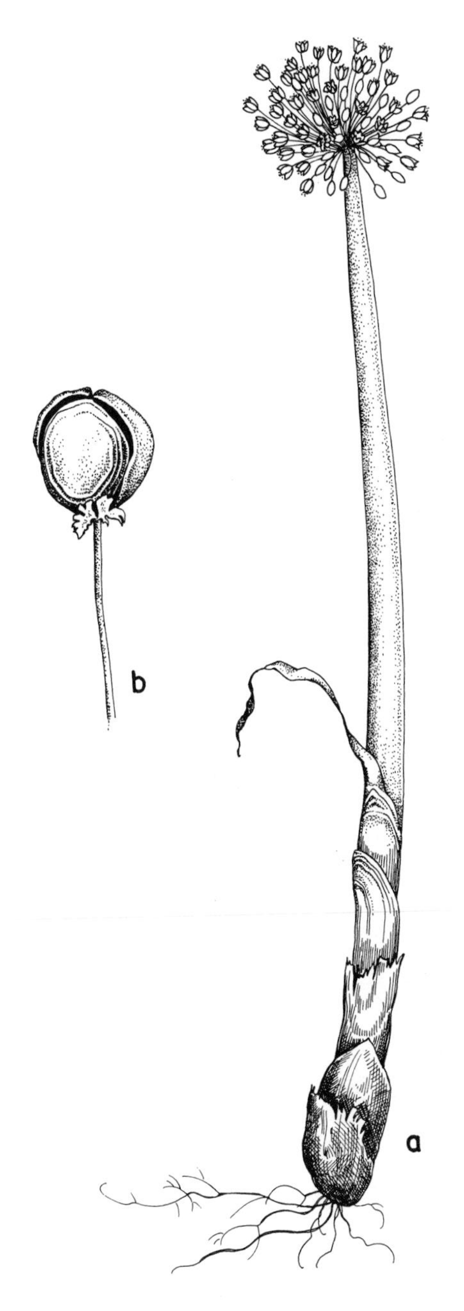

43. *Allium cepa* (Onion). *a.* Habit, X¼. *b.* Capsule, X1¼.

44. *Nothoscordum bivalve* (False Garlic). *a.* Habit (in flower), X¼. *b.* Flower, X1¼. *c.* Fruiting cluster, X¼. *d.* Capsule, X1¼.

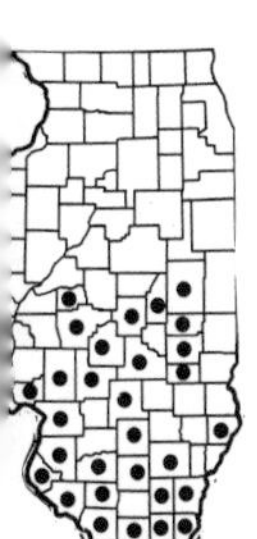

COMMON NAME: False Garlic.
HABITAT: Dry woods, bluffs, and prairies.
RANGE: Virginia to Kansas, south to Texas and Florida; West Indies; Mexico; Central America; South America.
ILLINOIS DISTRIBUTION: Rather common in the southern one-half of the state; apparently completely absent from the northern one-half. This species looks exactly like an onion or garlic, except that it lacks the typical pungent odor and bears 6–10 ovules per cell of the ovary. This species flowers in April, May, and June, and occasionally again in September and October.

23. *Medeola* L. – Cucumber Root

Inflorescence terminal, umbellate; flowers perfect; perianth parts 6, free; style 1, deeply 3-lobed; berry few-seeded; rhizomatous plants with two whorls of cauline leaves.

Only the following species comprises the genus.

1. **Medeola virginiana** L. Sp. Pl. 339. 1753. *Fig. 45.*

Rhizome white, slender, to 8 cm long; stem to 75 cm tall, erect, woolly when young, becoming glabrous at maturity; leaves oblong-oblanceolate, short-acuminate at apex, tapering or rounded at base, the lower in whorls of 5–11, the upper in whorls of 3–5; umbel terminal, sessile, solitary, 3- to 9-flowered; flowers greenish-yellow, the pedicels to 25 mm long, spreading or pendulous; perianth parts narrowly lanceolate, recurved, 7–8 mm long; berry dark purple, 4–8 mm in diameter; 2n = 14 (Stewart & Bamford, 1942).

COMMON NAME: Indian Cucumber Root.
HABITAT: Rich, moist woodlands.
RANGE: Nova Scotia to Minnesota, south to Louisiana and Florida.
ILLINOIS DISTRIBUTION: Rare; known from Cook and LaSalle counties.
This distinct plant, with its two whorls of leaves and its solitary umbel of pendulous flowers, blooms in May in Illinois. It may be confused in the sterile condition with a whorled-leaf orchid, *Isotria verticillata.*

The common name comes from the rhizome which has the flavor of a cucumber.

45. *Medeola virginiana* (Indian Cucumber Root). *a.* Habit (in flower), X⅙. *b.* Flower, X1. *c.* Habit (in fruit), X⅙. *d.* Capsule, X1. *e.* Seed, X2.

Only four collections have been made of this species in Illinois, all from rich moist woods. The first three were made in Cook County, beginning with Locy's collection in 1877 from West LaGrange. This was followed by a collection by L. N. Johnson from Evanston in 1889 and by Buhl from Edgebrook in 1916. Then Fuller made the remarkable discovery in 1939 of this species from Starved Rock State Park in LaSalle County.

24. *Trillium* L. – Wake Robin

Flower solitary, terminal, perfect; perianth parts 6, free, the outer three (sepals) green, the inner three (petals) white, maroon, yellow, or greenish; style none or 1 and minute; stigmas 3; berry rather dry, several-seeded; rhizomatous plants with a single whorl of leaves.

The species of this genus are highly variable. Flower parts and leaves occasionally may be in twos, fours, fives, or sixes. Note should be taken that these abnormal specimens will not key out properly. White-flowered species may have pink or maroon flowers; maroon-flowered species may have yellow, greenish, or whitish flowers. Considerable disagreement exists among taxonomists as to the delineation of species.

KEY TO THE TAXA OF Trillium IN ILLINOIS

1. Flower sessile, basically maroon or green (rarely yellow).
  2. Leaves petiolate; petals abruptly tapering to distinct claws; sepals reflexed ______ 1. *T. recurvatum*
  2. Leaves sessile; petals scarcely or gradually tapering, usually without distinct claws; sepals spreading or erect, not reflexed.
    3. Stamens one-half as long as petals; connectives 1–3 mm long.
      4. Stems and veins on lower leaf surface glabrous; anthers 5–6 times longer than the filaments; pollen yellow ______ 2. *T. sessile*
      4. Summit of stems and veins on lower leaf surface minutely scabrous; anthers 4 times longer than the filaments; pollen olive to brownish ______ 3. *T. viride*
    3. Stamens one-third as long as petals; connectives less than 1 mm long ______ 4. *T. cuneatum*
1. Flower pedunculate, white or occasionally pink or purple, never green or yellow.
  5. Ovary and fruit 3-angled ______ 5. *T. nivale*
  5. Ovary and fruit 6-angled.

6. Ovary purple (in Illinois specimens) ________ 6. *T. erectum*
6. Ovary white.
   7. Petals strongly recurved, 1.5–2.5 cm long; peduncle 10–40 mm long; anthers 4–7 mm long ________ 7. *T. cernuum* var. *macranthum*
   7. Petals spreading, 2–6 cm long; peduncle 25–100 mm long; anthers (6–) 7–15 mm long.
      8. Stigmas erect; anthers slightly longer than the filaments ________ 8. *T. grandiflorum*
      8. Stigmas spreading; anthers about twice as long as the filaments ________ 9. *T. flexipes*

1. **Trillium recurvatum** Beck, Am. Journ. Sci. 11:178. 1826. *Fig. 46.*

Trillium recurvatum f. *luteum* Clute, Am. Bot. 28:79. 1922.
Trillium recurvatum f. *shayii* Palmer & Steyerm. Ann. Mo. Bot. Gard. 22:504. 1935.

Stems to 40 (–65) cm tall, glabrous; leaves elliptic to lance-ovate, acute or subacute, petiolate at base, to 12 cm long, to 7 cm broad, mottled with purple; flower sessile; sepals reflexed, lanceolate, short-acuminate, to 3 cm long; petals erect, oblong-lanceolate to ovate, to 2.5 (–3.5) cm long, acute, tapering to a distinct claw at the base, maroon, rarely greenish-yellow or yellow; stamens maroon, rarely yellow, about half as long as the petals, the anther about three times longer than the filament, the connective 1–3 mm long; 2n = 10 (unpublished data).

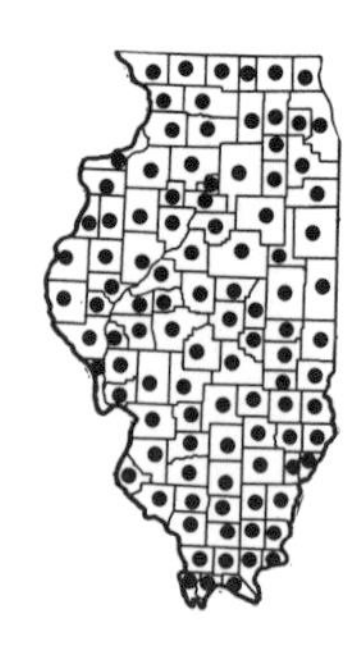

COMMON NAME: Red Trillium; Wake Robin.
HABITAT: Rich, moist woodlands.
RANGE: Michigan to Iowa, south to Louisiana and Alabama.
ILLINOIS DISTRIBUTION: Common; known from every county.

Although the majority of specimens have maroon petals and stamens, some variation may occur. Plants with yellowish petals and maroon stamens may be known as f. *luteum* and are found occasionally throughout the state. Plants with yellow petals and stamens may be called f. *shayii*, known from about six counties in the central and southern parts of the state. These variants are not distinguished on the distribution map, and might be expected in any population.

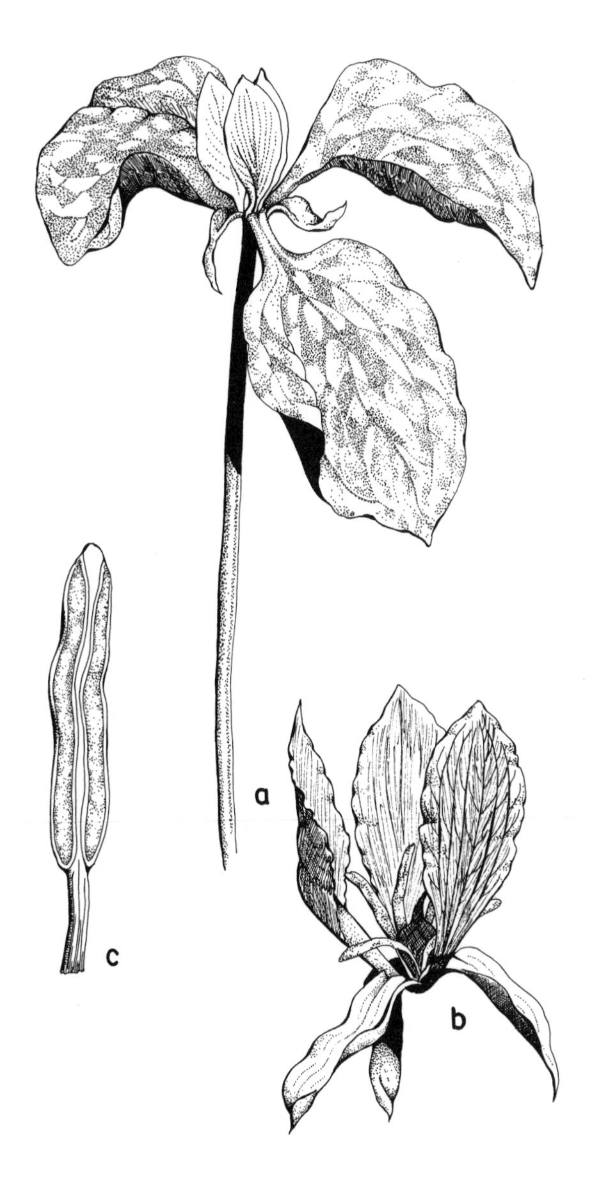

46. *Trillium recurvatum* (Red Trillium). *a*. Habit, X⅜. *b*. Flower, X1. *c*. Stamen, X3¾.

Flowers have been observed in specimens from the southern counties as early as the last week in March in years following a rather mild winter.

Trillium recurvatum is the only maroon-flowered species in the eastern United States with petiolate leaves.

Variation is known in the number of floral parts and leaves.

2. **Trillium sessile** L. Sp. Pl. 340. 1753. *Fig. 47.*

Trillium sessile f. *viridiflorum* Beyer, Torreya 27:83. 1927.

Stems to 30 cm tall, glabrous; leaves oval to nearly orbicular, obtuse or abruptly short-tipped at the apex, rounded at the sessile base, to 7 (–9) cm long, nearly as broad, sparsely mottled with purple; flower sessile; sepals spreading or ascending, lance-ovate, to 2.5 (–3.0) cm long; petals erect, elliptic to lance-ovate, acute or obtuse, tapering gradually to the usually clawless base, to 3 (–4) cm long, maroon or greenish; stamens maroon, half as long as the petals, the anther 5–6 times longer than the filament, the connective 1–3 mm long; capsule usually globoid or subgloboid, 0.9–1.3 cm long, nearly as broad.

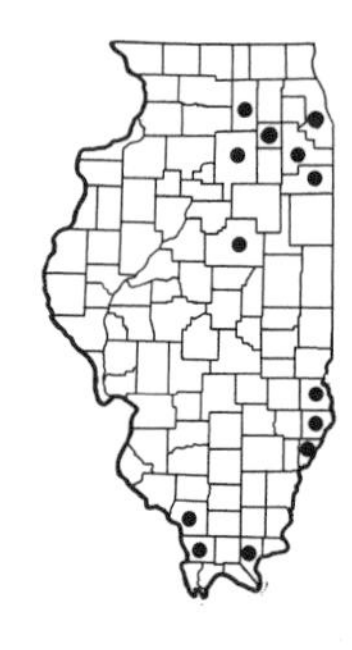

COMMON NAME: Sessile Trillium; Sessile Wake Robin.

HABITAT: Rich woodlands.

RANGE: New York to Kansas, south to Arkansas, Virginia, and extreme northern Alabama.

ILLINOIS DISTRIBUTION: Occasional; scattered throughout the state, except in the western counties.

This species is generally the smallest of the sessile-flowered species. Although a yellow-green color-form is known throughout its range, this variant apparently has not been found in Illinois. The shorter, more ovate petals distinguish this species from either *T. viride* or *T. cuneatum.* Another useful character in distinguishing this species from *T. viride* is the completely glabrous undersurface of the leaves in *T. sessile.*

Trillium sessile, under favorable conditions, may begin to flower in late March.

3. **Trillium viride** Beck, Am. Journ. Sci. 11:178. 1826. *Fig. 48.*

Stems to 45 cm tall, glabrous except for the minutely scabrous apex; leaves narrowly ovate, acute or rounded at the apex, more or less rounded at the base, to 8 cm long, to 5.5 cm broad, oc-

47. *Trillium sessile* (Sessile Trillium). *a.* Habit (in flower), X¼. *b.* Flower, X1. *c.* Stamen, X2½. *d.* Habit (in fruit), X¼.

48. *Trillium viride* (Green Trillium). *a.* Habit, X¼. *b.* Flower, X⅝. *c.* Stamen, X2.

casionally mottled with purple, scabrous below on the veins; flower sessile; sepals spreading or ascending, lance-elliptic, to 3 cm long; petals green (rarely maroon), erect, lance-elliptic, gradually tapering to the usually clawed base, to 6 cm long; stamens maroon, about one-half as long as the petals, the anther about four times longer than the filament, the connective a little more than 1 mm long; capsule usually ovoid, 1.3–2.2 cm long, nearly as broad.

COMMON NAME: Green Trillium.

HABITAT: Rich woodlands; prairies.

RANGE: Endemic in eastern Missouri and adjacent counties of southwestern Illinois.

ILLINOIS DISTRIBUTION: Not common; scattered in the southern two-thirds of the state, apparently a little more common in the more southern counties.

The majority of specimens are green-flowered. The hirtellous upper stem and lower leaf surfaces along with the tapering green petals help distinguish this species. Maroon-flowered specimens have been recorded for this species.

The following comments concerning *T. viride* and *T. cuneatum* have been sent to me in a personal communication from Dr. John D. Freeman, Auburn University:

A definitive feature of *T. viride* is the presence of numerous stomates, uniformly distributed, in the upper epidermis of the bracts ("leaves"). The stomates give the upper surface an almost farinose appearance and show up as tiny whitish dots in wilted or press-dried specimens. No stomates at all were found in *T. cuneatum*. The feature was constant and characteristic for all plants of *T. viride*, except those from the Giant City Park area. Some plants from there have stomates, others none. In fact, [*Mohlenbrock*] *14214* has none, and I can find no good reason why it shouldn't be called *T. cuneatum* as far as its morphology is concerned.

Dr. Freeman goes on to point out, however, that it is likely that *Mohlenbrock 14214* is a *T. viride* without the usual stomates, but that it is disturbingly cuneatum-like.

4. **Trillium cuneatum** Raf. Aut. Bot. 133. 1840. *Fig. 49.*

Trillium hugeri Small, Fl. S. E. U. S. 277. 1903.

Stems 35 cm tall (in Illinois), glabrous; leaves broadly ovate to orbicular, short-acuminate at apex, more or less rounded at

49. *Trillium cuneatum* (Trillium). *a.* Habit X¼. *b.* Flower, X¾.

base, to 12 cm long, to 10 cm broad, mottled with purple; flower sessile; sepals spreading or ascending, lance-elliptic, subacute, to 5 cm long; petals maroon or greenish, erect, lance-elliptic, scarcely tapering to the base, to 6 cm long; stamens maroon, about one-third as long as the petals, the anther about four times longer than the filament, the connective less than 1 mm long.

COMMON NAME: Trillium.

HABITAT: Rich, moist woodlands.

RANGE: North Carolina to Illinois, south to Mississippi and Georgia.

ILLINOIS DISTRIBUTION: Very rare; known only from Jackson County (Giant City State Park, April 28, 1960, *R. H. Mohlenbrock 14214*, and again in 1968).

This species may be distinguished from maroon-flowered plants of *T. viride* by its completely glabrous stems and leaves, its scarcely tapering petals, its shorter stamens in relation to the petals, and its very short connectives.

Trillium cuneatum is known in Illinois only from a single locality.

Trillium hugeri Small is the same species.

5. **Trillium erectum** L. Sp. Pl. 340. 1753. *Fig. 50.*

Stems to 60 cm tall, glabrous; leaves broadly ovate to suborbicular, acuminate, rounded at the sessile base, to 20 cm long, nearly as broad; peduncle usually straight and erect, rarely reflexed, to 10 cm long, glabrous; sepals lanceolate, acuminate, to 3 (–5) cm long; petals purple (in Illinois specimens), spreading from the base, narrowly ovate-oblong, acute or acuminate, to 3 (–5) cm long; anthers less than twice as long as the filaments.

COMMON NAME: Purple Trillium.

HABITAT: Rich woods.

RANGE: Gaspé Peninsula to Ontario, south to northern Illinois, Georgia, and Delaware.

ILLINOIS DISTRIBUTION: Known only from two collections from McHenry County; also Lake County.

Apparently the original collection of this species from Illinois was made by George Vasey from near Ringwood, McHenry County, in the nineteenth century. This collection has been reported previously as *T.*

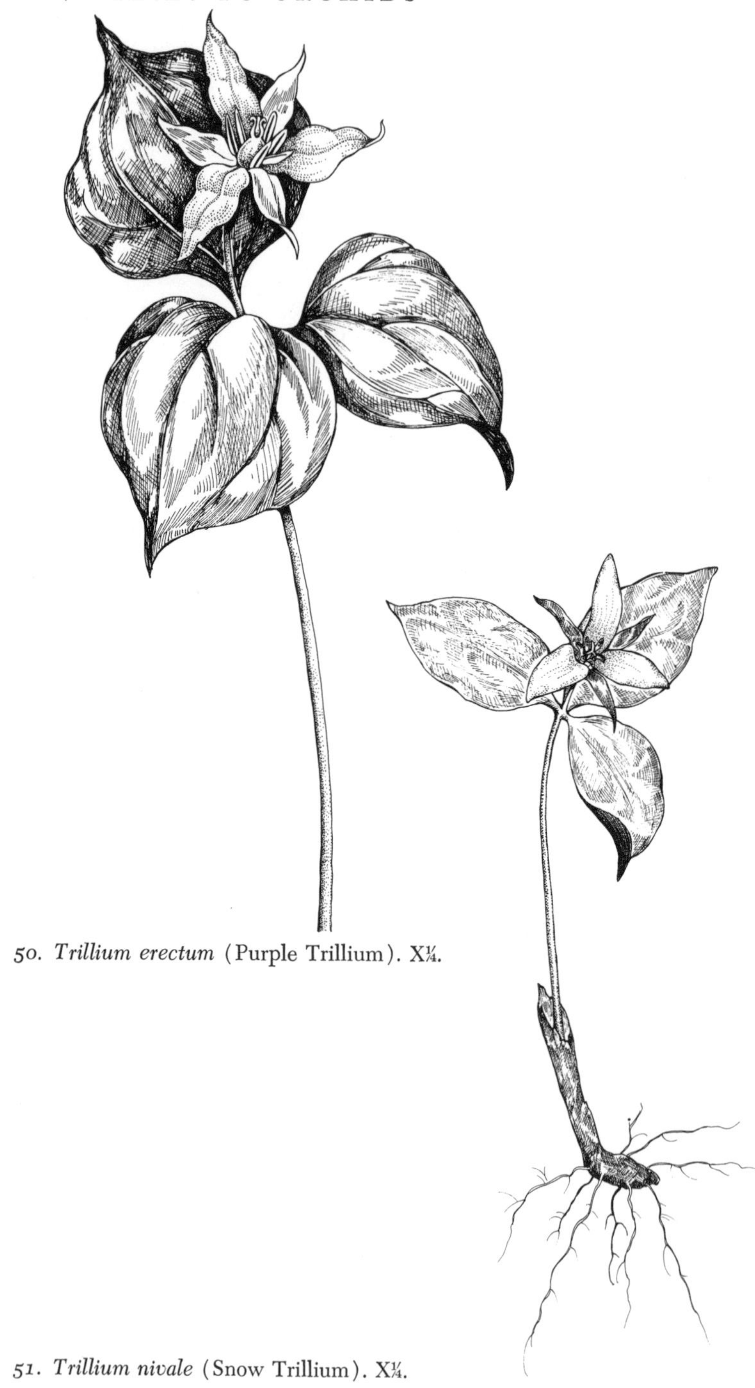

50. *Trillium erectum* (Purple Trillium). X¼.

51. *Trillium nivale* (Snow Trillium). X¼.

cernuum var. *macranthum*, but the specimen in question seems to have a purple-brown ovary. A more recent collection of *T. erectum* was made on May 9, 1954, by Julian A. Steyermark from a rich wooded slope along the headwaters of a tributary of Boone Creek, T 44 N, R 7 E, section 11, on the property of Mr. Richard Babcock, four miles east of Woodstock. Associated species were *Trillium grandiflorum*, *Hepatica acutiloba*, *H. americana*, *Mitella diphylla*, and *Asclepias exaltata*.

5. **Trillium nivale** Riddell, Syn. Fl. W. States 93. 1835. *Fig. 51.*

Stems to 15 cm tall, glabrous; leaves elliptic to ovate, subacute or obtuse at the apex, rounded at the base, to 5 cm long at anthesis, enlarging to 10–12 cm long later, petiolate; petiole glabrous, 5–10 cm long; peduncle erect at first, later arching, glabrous, to 3 cm long; sepals lanceolate, short-acuminate, to 2 cm long, petals white, usually pinkish near base, spreading, broadly elliptic, short-tipped, to 3 (–4) cm long; anthers a little longer than the slender filaments; capsule 0.7–1.0 cm long.

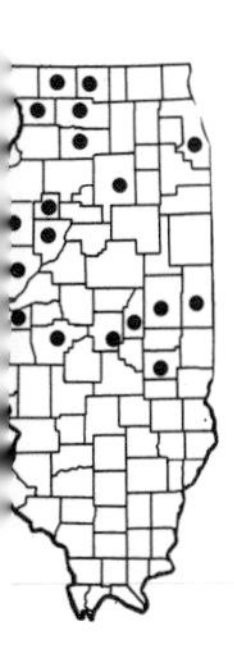

COMMON NAME: Snow Trillium.

HABITAT: Rich woodlands.

RANGE: Pennsylvania to Minnesota, south to Missouri and Kentucky.

ILLINOIS DISTRIBUTION: Local; restricted to the northern three-fifths of Illinois.

The common name is derived from the early flowering of this species in March while snow is often on the ground. It is the only pedunculate species with strongly petiolate leaves. It is the smallest of the Illinois species of *Trillium*.

6. **Trillium cernuum** L. var. **macranthum** Eames & Wieg. Rhodora 25:191. 1923. *Fig. 52.*

Stems to 45 cm tall, glabrous; leaves broadly ovate to suborbicular, short-acuminate, rounded or more or less tapering to an obscure petiole, to 12 (–15) cm long; peduncle recurved, 2–4 cm long, glabrous; sepals lanceolate, short-acuminate, 1.5–2.5 cm long; petals white, recurved at the tips, obovate, obtuse or subacute, 1.5–2.5 cm long; anthers about twice as long as the flattened filaments.

COMMON NAME: Nodding Trillium.
HABITAT: Moist woodlands.
RANGE: Vermont to Mackenzie, south to Iowa, Tennessee, and Pennsylvania.
ILLINOIS DISTRIBUTION: Very rare; known only from the original collection from Wolf Lake, Cook County, by E. J. Hill in 1891. The report of this species from McHenry County, based on a Vasey collection from near Ringwood, is based on a misidentification for *T. erectum*.

7. **Trillium grandiflorum** (Michx.) Salisb. Parad. Lond. 1:pl. 1. 1806. *Fig.* 53.

Trillium rhomboideum var. *grandiflorum* Michx. Fl. Bor. Am. 1:216. 1803.

Trillium grandiflorum var. *parvum* R. R. Gates, Ann. Mo. Bot. Gard. 4:58. 1917.

Stem to 35 (–40) cm tall, glabrous; leaves ovate to suborbicular, short-acuminate, rounded or tapering to the usually sessile base, up to 12 cm long at anthesis, expanding later; peduncle ascending, 2–8 (–10) cm long, glabrous; sepals lanceolate, obtuse or subacute, spreading, 3–5 cm long; petals white, sometimes becoming pink-tinged through aging, obovate, spreading, acute, 4–6 cm long; stigmas nearly erect; anthers slightly longer than the filaments; 2n = 10 (Darlington & LaCour, 1940).

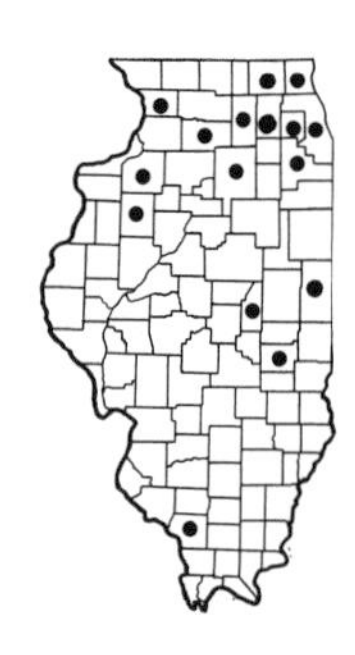

COMMON NAME: Large White Trillium.
HABITAT: Rich, moist woodlands.
RANGE: Quebec to Minnesota, south to Iowa, southern Illinois, and Georgia.
ILLINOIS DISTRIBUTION: Occasional in the northern half of the state; absent from the southern half except for a single station in Jackson County, from Little Grand Canyon, near Murphysboro.

This is the largest-flowered and most handsome member of the genus in Illinois. It usually reaches proportions greater than that of *T. flexipes*, a species with recurved stigmas, more elongate anthers in ratio to the filaments, and petals about the same length as the sepals.

Much variation occurs throughout the range of this species, although few variants have been collected in Illinois.

The flowers are produced in late April and May.

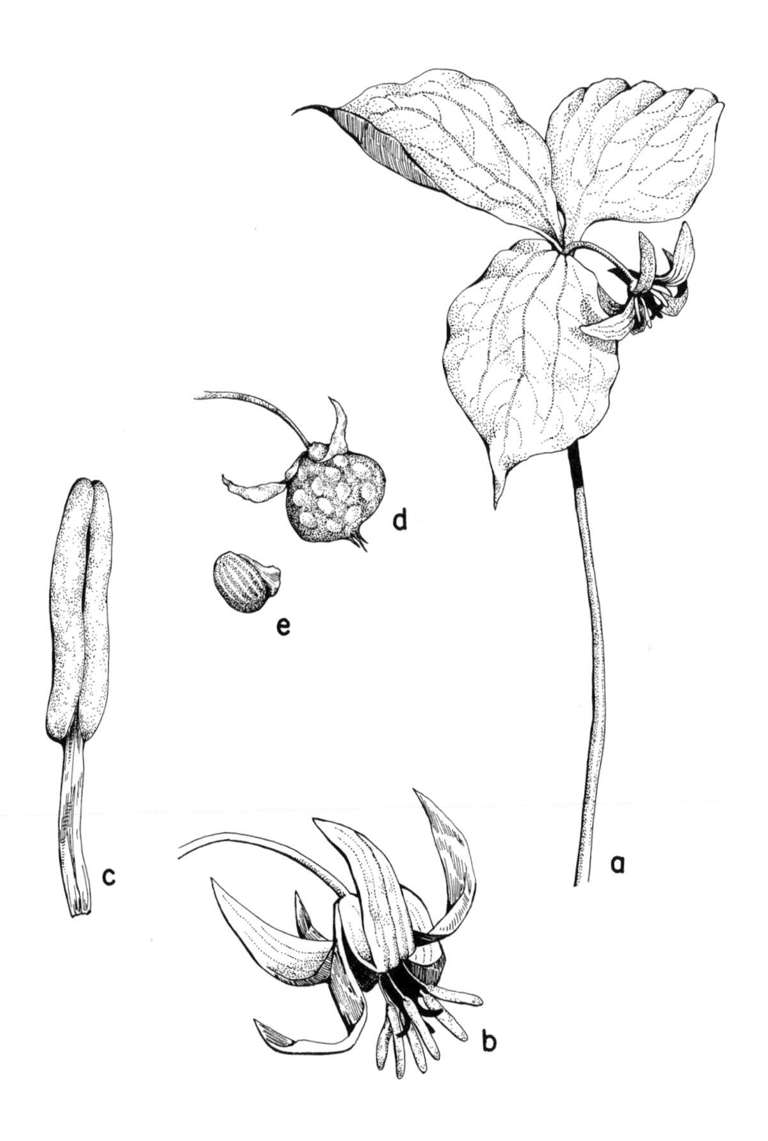

52. *Trillium cernuum* var. *macranthum* (Nodding Trillium). *a.* Habit, X¼. *b.* Flower, X¾. *c.* Stamen, X2½. *d.* Capsule, X½. *e.* Seed, X½.

53. *Trillium grandiflorum* (Large White Trillium). *a.* Habit, X¼. *b.* Flower (all but one perianth part removed), X¾. *c.* Stamen, X2. *d.* Habit (in fruit), X¼. *e.* Seed, X½.

8. Trillium flexipes Raf. Aut. Bot. 133. 1840. *Fig. 54.*

Trillium erectum var. *declinatum* Gray, Man. Bot. 523. 1817.
Trillium declinatum (Gray) Gleason, Bull. Torrey Club 33:389. 1906, non Raf. (1840).
Trillium gleasoni Fern. Rhodora 34:21. 1932.

Stem to 50 cm tall, glabrous; leaves broadly rhombic, acute to short-acuminate, rounded or tapering to the sessile base, to 15 cm long; peduncle ascending or pendulous, 2–10 cm long, glabrous; sepals lanceolate, acute to short-acuminate, 2–5 cm long;

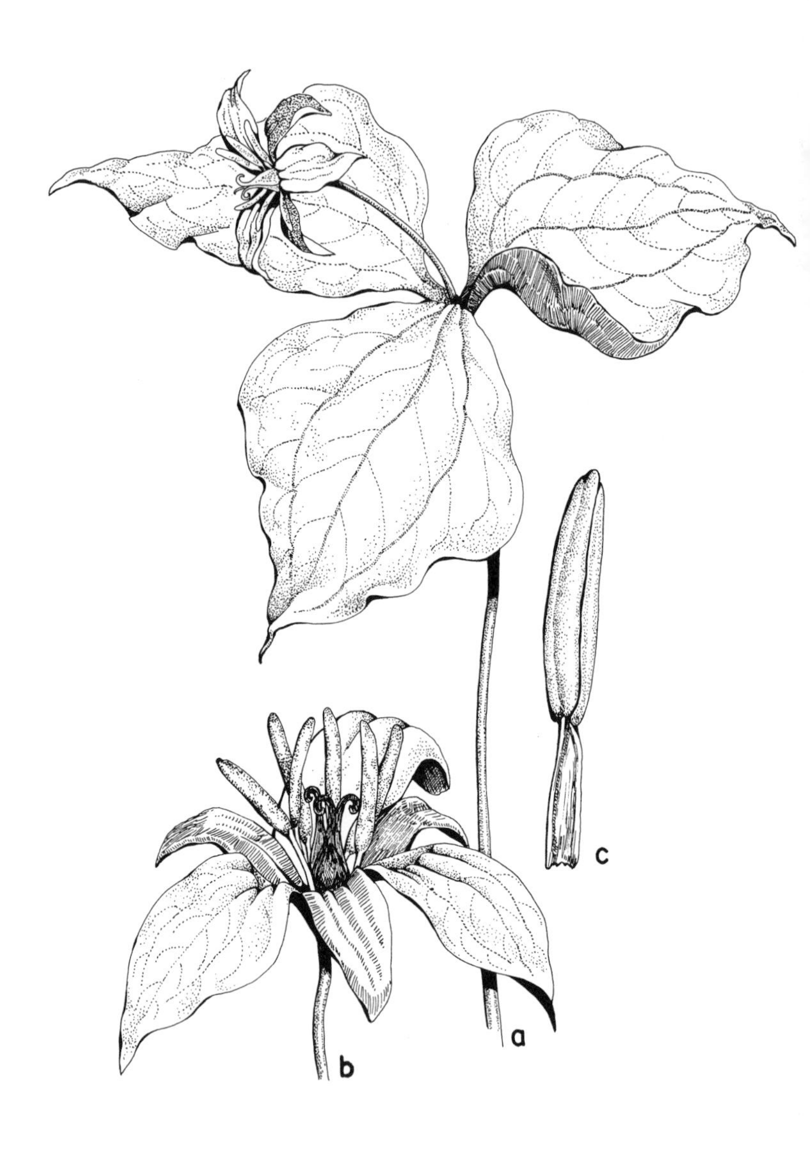

54. *Trillium flexipes* (White Trillium). *a.* Habit, X¼. *b.* Flower, X⅝. *c.* Stamen, X5.

petals white rarely pinkish-tinged, lance-ovate, obtuse to subacute, spreading, 2–5 cm long; stigmas recurved; anthers nearly twice as long as the filaments; 2n = 10 (unpublished data).

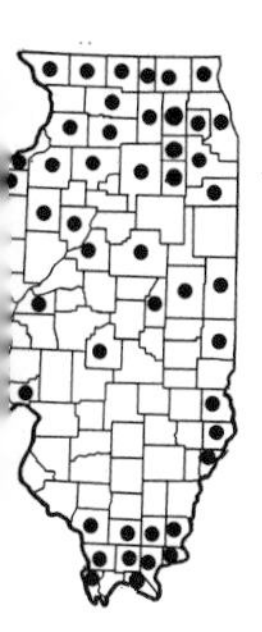

COMMON NAME: White Trillium.
HABITAT: Rich woodlands.
RANGE: New York to Minnesota, south to Missouri, Tennessee, and Maryland.
ILLINOIS DISTRIBUTION: Occasional throughout the state; absent from the interior south-central counties. This is the most common white-flowered *Trillium* in Illinois, flowering during the last of April and most of May. There is some question concerning the exact species meant by Rafinesque in his description of *T. flexipes*. Those wishing to disregard his binominal use *T. gleasonii*.

A colony of a dozen plants at Giant City State Park (Jackson County) has been discovered in which the petals number four, five, or six. In addition, there may be only five stamens, although six is the usual number in this variation.

25. *Yucca* L. – Adam's Needle

Inflorescence paniculate (in Illinois plants); flowers perfect; perianth parts 6, free, glandless; style 1; capsule loculicidal; evergreen, fleshy plants from a stout, woody caudex.

Only the following escaped taxon occurs in Illinois.

1. **Yucca filamentosa** L. var. **smalliana** (Fern.) Ahles, Journ. Elisha Mitch. Sci. Soc. 80 (1):172. 1964. *Fig.* 55.

Yucca smalliana Fern. Rhodora 46:8. 1944.

Leaves basal, rather fleshy, numerous, oblanceolate, acute and often spine-tipped at apex, to 75 cm long, to 7.5 cm broad, keeled along the midvein, fibrous and roughened along the margins; inflorescence paniculate, to nearly 3 m long; perianth parts 5–7 cm long, white to cream, broadly ovate, short-acuminate, broadly rounded at base, glandless; style 8–10 mm long; capsule broadly cylindric, constricted near the middle, to 4 cm long; seeds rounded, flat, 6–7 mm long, black, shiny; 2n = 60 (McKelvey & Sax, 1933, as *Y. filamentosa*).

55. *Yucca filamentosa* var. *smalliana* (Adam's Needle). *a.* Habit (shaded), X1/60. *b.* Flowers, X1/4. *c.* Leaf, X1/8.

COMMON NAME: Yucca; Adam's Needle.
HABITAT: Waste ground, along roads, in cemeteries.
RANGE: Occurs naturally only from New Jersey to Georgia; widely planted and often escaped from cultivation.
ILLINOIS DISTRIBUTION: Scattered throughout the state; abundant in cemeteries.
The large fleshy basal leaves with fibrous margins make this plant easy to identify during any season.

This species, a common escape from cultivation, has been called *Y. filamentosa* by previous Illinois authors, but I am following several recent authors in regarding typical *Y. filamentosa* as a species restricted to the Atlantic Coastal Plain. Our plant does not appear to merit specific recognition. It is here known as *Y. filamentosa* var. *smalliana*.

26. *Leucojum* L. – Snowflake

Only the following species occurs in Illinois.

1. Leucojum aestivum L. Sp. Pl. 975. 1753. *Fig. 56.*

Scapose perennial from a cluster of tunicated bulbs; leaves sword-shaped, glabrous, to 1.5 cm broad; scape more or less erect, glabrous, to 75 cm tall, bending downward in fruit; flowers 2-several, nodding without a corona, borne on elongated pedicels; perianth campanulate, the six segments mostly equal, white with green tips, to 20 mm long; stamens 6, borne at the base of the perianth segments; capsule obovoid.

COMMON NAME: Summer Snowflake.
HABITAT: Roadside ditch (in Illinois).
RANGE: Native in Europe; occasionally escaped from cultivation in the United States.
ILLINOIS DISTRIBUTION: Known only from Pope County.
The summer snowflake is readily distinguished by its campanulate flowers with green-tipped perianth segments. This species is occasionally grown as an ornamental in Illinois but rarely escapes. At the Illinois station, several specimens occur. The flowers appear in late April.

56. *Leucojum aestivum* (Summer Snowflake). X¼.

27. *Narcissus* L. – Narcissus

Perennial herbs from bulbs; leaves linear, basal, more or less fleshy; inflorescence of a solitary flower (in the escaped Illinois species) or umbellate and few-flowered; flowers recurved; perianth parts 6, generally uniform in color, short-tubular below, spreading above; corona between spreading perianth parts and stamens; ovules several per cell of ovary.

This is a commonly cultivated genus in Illinois, with several species used ornamentally. Only the following two species have been collected as escapes in Illinois.

KEY TO THE SPECIES OF Narcissus IN ILLINOIS

1. Perianth yellow; corona yellow, tubular, about as long as the perianth ---------------------------------- 1. *N. pseudo-narcissus*
1. Perianth white; corona white, with red margin, cupular, less than one-fourth as long as the perianth ---------------- 2. *N. poeticus*

1. Narcissus pseudo-narcissus L. Sp. Pl. 289. 1753. *Fig.* 57.

Scapose perennial from bulbs; scapes to 45 cm tall; blades basal, flat, linear-elongated, obtuse to subacute, glaucous, 30–45 cm long, 1–2 cm broad; flower solitary, 4–6 cm long, the tube broad, 5–12 mm long, the perianth segments oblong to ovate, the corona 2.5–3.5 cm long, more or less undulate on the edge, with all parts usually some shade of yellow.

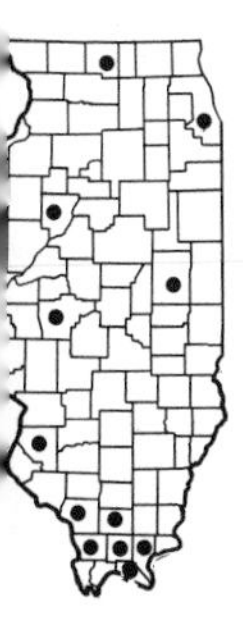

COMMON NAME: Daffodil.

HABITAT: Abandoned homesteads; along roads.

RANGE: Native of Europe; occasionally escaped from cultivation.

ILLINOIS DISTRIBUTION: Scattered throughout the state.

This species persists and spreads easily around old homesteads.

2. Narcissus poeticus L. Sp. Pl. 289. 1753. *Fig.* 58.

Scapose perennial from bulbs; scapes to 50 cm tall; blades basal, flat, linear-elongate, obtuse, glaucous, 30–45 cm long, 0.5–1.5 cm broad; flower solitary, fragrant, the tube rather broad, 18–25 mm long, white, the perianth segments broadly obovate, apiculate, overlapping, white, the corona very short, white, with a red, crisped edge.

57. *Narcissus pseudo-narcissus* (Daffodil). X¼.

58. *Narcissus poeticus* (Poet's Narcissus). X¼.

COMMON NAME: Poet's Narcissus.
HABITAT: Along roads.
RANGE: Native of Europe; rarely escaped from cultivation.
ILLINOIS DISTRIBUTION: Escaped specimens have been collected in several Illinois counties.
Variation occurs in the length and color of the corona in this species which persists around old homesteads.

28. *Hymenocallis* SALISB. – Spider Lily

Perennial herbs from bulbs; leaves broadly linear, basal, fleshy; inflorescence umbellate; flowers ascending; perianth parts 6, uniform in color, long-tubular below, spreading and often recurved above; corona connecting the filaments; ovules 2 per cell of ovary; capsule few-seeded.

There is an upwardly extending outgrowth of the perianth between the inner whorl of perianth parts and the stamens, called the corona, or crown.

Only the following species occurs in Illinois.

1. **Hymenocallis occidentalis** (LeConte) Kunth, Enum. 5:856. 1850. *Fig.* 59.

Pancratium occidentale LeConte, Ann. Lyc. New York 3:146. 1830.

Leaves broadly linear, to 60 cm long, to 4.5 cm broad, glabrous, glaucous; scape to 75 cm tall, glabrous, glaucous; inflorescence solitary, umbellate, 2- to 6-flowered; flowers white, ascending; bracts 2-several, linear-lanceolate, to 5 cm long; perianth tubular below, widely spreading above, the tube 5–10 cm long, the free segments linear, to 8 (–10) cm long; corona membranous, funnelform, 2.5–3.5 cm long, erose; filaments white, connected by the corona, nearly twice as long as the corona; capsule few-seeded.

59. *Hymenocallis occidentalis* (Spider Lily). *a.* Inflorescence, X¼. *b.* Leaf, X¼.

COMMON NAME: Spider Lily.
HABITAT: Low woods, frequently in swampy situations.
RANGE: Indiana to Missouri, south to Alabama and Georgia.
ILLINOIS DISTRIBUTION: Not common; confined to the extreme southern counties; also Wabash County.

This is one of the most beautiful wildflowers in Illinois. The delicate perianth, whose slender, spreading, free segments account for the common name, expands in late July. Flowers may open as late as mid-September.

29. *Polianthes* L. – Aloe

Perennial herbs from a crown of thick, fibrous roots; leaves basal, fleshy; inflorescence spicate, many-flowered; perianth parts 6, tubular, uniform in color; ovary inferior; ovules several per cell of ovary; capsule loculicidal, many-seeded.

In addition to the following species which is native in southern Illinois, the tuberose, *P. tuberosa* L., is frequently cultivated in the area.

1. **Polianthes virginica** (L.) Shinners, Sida 2(4):335. 1966. *Fig. 60.*

Agave virginica L. Sp. Pl. 323. 1753.
Manfreda virginica (L.) Salisb. Gen. Pl. Fragm. 78. 1866.
Agave virginica var. *tigrina* Engelm. Trans. Acad. Sci. St. L. 3:302. 1875.
Manfreda tigrina (Engelm.) Small, Fl. SE. U. S. 287. 1903.
Agave virginica f. *tigrina* (Engelm.) Palmer & Steyerm. Ann. Mo. Bot. Gard. 22:507. 1935.
Agave tigrina (Engelm.) Cory, Rhodora 38:405. 1936.
Polianthes virginica f. *tigrina* (Engelm.) Shinners, Sida 2(4):335. 1966.

Leaves lanceolate to oblanceolate, to 40 cm long, to 5 cm broad, serrulate (rarely entire), occasionally purple-flecked; scape erect, glabrous, to nearly 2 m tall; inflorescence spicate, unbranched, many-flowered; perianth tubular below, spreading above, cream to greenish, the tube 12–17 mm long, the lobes 5–8 mm long; capsule subgloboid, to 22 mm long, to 16 mm broad, short-beaked.

60. *Polianthes virginica* (American Agave). *a.* Habit (shaded), X1⁄60. *b.* Inflorescence, X¼. *c.* Capsules, X⅜. *d.* Seed, X1¼. *e.* Leaf, X¼.

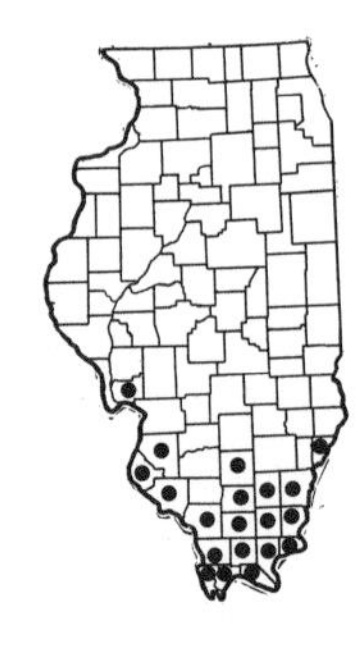

COMMON NAME: American Agave; American Aloe; False Aloe.

HABITAT: Sandstone outcroppings; dry woodlands.

RANGE: Virginia to Missouri, south to Texas and Florida.

ILLINOIS DISTRIBUTION: Occasional; restricted to the southern one-fourth of the state; also Jersey County.

This is one of the desert-looking species which grows on sandstone ledges of southern Illinois. The flowers, which are mildly fragrant in the evening, appear the last of May and continue to the first of July. Specimens with purple-flecked leaves have been designated f. *tigrina.*

Originally described by Linnaeus as an *Agave,* this species later was segregated into the genus *Manfreda* by Salisbury on the basis of the spineless leaves and the flowers borne one at a node in the simple inflorescence. True *Agave,* on the other hand, is characterized by persistent, spine-tipped leaves and the flowers borne in pairs at a node in the usually paniculate inflorescence. Shinners (1966) further points out that no suitable differences can be found to separate *Manfreda* from the genus *Polianthes.* If one, therefore, desires to break-up *Agave* in the manner mentioned above and to consider *Manfreda* and *Polianthes* as indistinct from each other, then the prior genus name *Polianthes* must be used for our species in Illinois.

30. *Hypoxis* L. – Star Grass

Perennial herbs from very short rhizomes; leaves linear, basal, not fleshy; inflorescence more or less umbellate, with 2–6 flowers; perianth parts 6, uniform in color, divided to the ovary; ovary inferior; ovules several per cell of ovary; fruit a capsule.

Only the following species occurs in Illinois.

1. Hypoxis hirsuta (L.) Coville, Mem. Torrey Club 5:118. 1894. *Fig. 61.*

Ornithogalum hirsutum L. Sp. Pl. 306. 1753.

Hypoxis erectum L. Syst. 2:986. 1759.

Rhizomes very short, corm-like, to 2 cm thick; leaves linear, to 50 cm long, to nearly 1 cm broad, entire, pilose; scape ascending, sparsely pilose, to 30 cm tall; inflorescence more or less umbellate, with 2–6 flowers; pedicels sparsely pilose, to 25 mm long; perianth parts yellow and glabrous above, greenish and

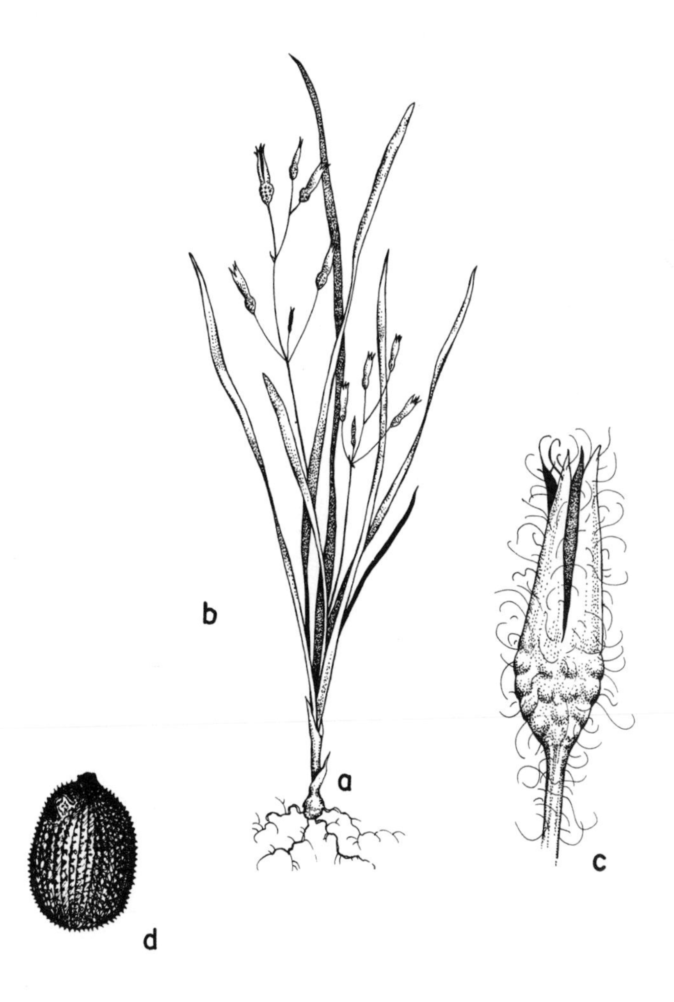

61. *Hypoxis hirsuta* (Yellow Star Grass). *a.* Habit, X¼. *b.* Flower, X1¼. *c.* Capsule, X1¼. *d.* Seed, X10.

minutely pilose below, lanceolate to lance-ovate, subacute to acute, to 12 mm long; capsule ellipsoid, sparsely pilose, indehiscent, 3–6 mm long, crowned with the beak-like remains of the perianth; seeds minutely spiny, black, shining, 1.0–1.5 mm long.

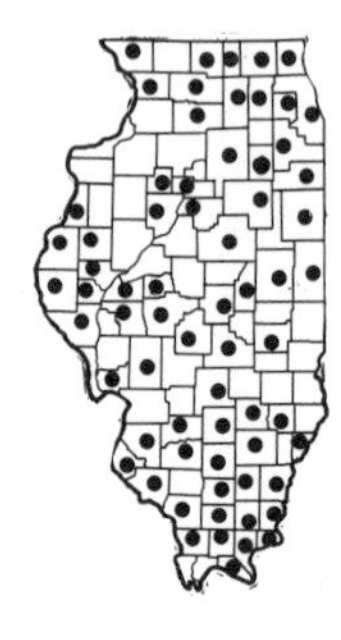

COMMON NAME: Yellow Star Grass.

HABITAT: Dry woods; prairies; fields; sandstone outcroppings; calcareous fens.

RANGE: Maine to Manitoba, south to Texas and Florida.

ILLINOIS DISTRIBUTION: Occasional in all parts of Illinois. The range for flowering of this species in Illinois is early April to mid-June.

In the northern half of Illinois, yellow star grass is primarily a prairie plant, occurring with *Comandra richardsiana, Krigia biflora, Dodecatheon meadia, Eryngium yuccifolium, Gentiana puberula, Heuchera richardsonii, Lithospermum canescens, Lobelia spicata, Pedicularis canadensis, Phlox pilosa, Senecio pauperculus, Silphium laciniatum, S. terebinthinaceum, Sporobolus heterolepis, Viola papilionacea, Zizia aurea,* and others. In calcareous fens it is associated with *Aster novae-angliae, Lilium michiganense, Oxypolis rigidior, Parnassia glauca, Smilacina stellata, Solidago ohioensis,* and *Valeriana ciliata.*

Linnaeus' treatment of this species was rather erratic. He first placed it in *Ornithogalum,* then six years later made it the type of his new genus, *Hypoxis,* but failed to use his previous epithet.

SMILACACEÆ – CATBRIER FAMILY

Twining or erect shrubs or herbs; leaves alternate (in Illinois) or opposite, simple, reticulate-veined; inflorescence often umbelliform; flowers unisexual and dioecious (in Illinois) or bisexual; perianth segments 6, uniform; stamens 6; ovary superior; fruit a berry.

Only the following genus occurs in Illinois.

1. *Smilax* L. – Catbrier

Twining (rarely erect) shrubs or herbs, often with tendrils; stems prickly in the woody species, terete or quadrangular;

leaves alternate, simple, reticulate-nerved, entire, sometimes lobed at base; inflorescence axillary, umbelliform; flowers dioecious, radially symmetrical; staminate flowers with 3 outer and 3 inner, nondistinguishable perianth segments and 6 stamens; pistillate flowers with 3 outer and 3 inner, nondistinguishable perianth segments, 1–6 staminodia, a single, trilocular, superior ovary, and 1–3 stigmas; fruit a berry, 1- to 3- (5-) seeded.

The tendrils are terminations of the stipules.

The Illinois representatives of this genus fall into two distinct groups. Section Smilax, the catbriers, is composed of the woody, prickly-stemmed species, while Section Coprosmanthus (=Section Nemexia), the carrion flowers, contains species which are herbaceous and unarmed.

Mangaly (1968) has revised Section Coprosmanthus, and his work is closely followed here.

KEY TO THE SPECIES OF Smilax IN ILLINOIS

1. Stems woody, with few to many prickles, rarely without prickles; ovule 1 per cell of ovary. Section Smilax.
    2. Leaves pale beneath, usually glaucous___________1. S. *glauca*
    2. Leaves green on both sides.
        3. Stems flexuous, with stout spines, or spines absent; leaves subcoriaceous to coriaceous.
            4. Leaves thick-margined, coriaceous, often blotched with white, often panduriform; peduncles much longer than the subtending petioles; berries not glaucous__2. S. *bona-nox*
            4. Leaves thin-margined, subcoriaceous, not blotched with white, never panduriform; peduncles about as long as the subtending petioles; berries glaucous___3. S. *rotundifolia*
        3. Stems not flexuous, with weak spines, or spines absent; leaves membranous_______________________________4. S. *hispida*
1. Stems herbaceous, without prickles; ovules 2 per cell of ovary. Section Coprosmanthus.
    5. Stems twining or climbing, with numerous tendrils all along the stem; peduncles borne from the axils of developed leaves.
        6. Leaves glabrous and glaucous beneath______5. S. *herbacea*
        6. Leaves pubescent beneath.
            7. Blades light green, not shiny beneath; petiole generally short; berries blue__________________6. S. *lasioneuron*
            7. Blades dark green, shiny beneath; petiole long; berries black______________________________7. S. *pulverulenta*

5. Stems erect, with few (or no) tendrils near the apex; peduncles borne from bladeless sheaths.
   8. Basal leaves narrowly ovate or elliptical, the base mostly truncate to subcordate; petioles usually equal to or longer than blade_____________________________8. *S. illinoensis*
   8. Basal leaves broadly ovate, the base cordate; petioles generally equal to or shorter than blade__________9. *S. ecirrata*

1. **Smilax glauca** Walt. Fl. Carol. 245. 1788. *Fig. 62.*

Climbing shrub from thick rhizomes; stems terete or subquadrangular, rather slender, often glaucous, with slender, firm prickles; leaves ovate, orbicular, elliptic, or subdeltoid, obtuse to subacute at apex, rounded or subcordate at base, coriaceous, green above, pale and usually glaucous beneath, glabrous or minutely pubescent beneath, entire, to 8 cm long, to 6 cm broad, 3- to 7-nerved; peduncles slender, arching or pendent, to 4 cm long, much exceeding the subtending petioles; berries bluish-black, glaucous, 4–5 mm in diameter.

This is one of the most distinct woody species of *Smilax* because of the glaucous lower leaf surface. Although it is relatively common in southern Illinois, it has not been found north of St. Clair and Lawrence counties. It is particularly common on exposed sandstone cliffs of the Shawneetown Ridge.

Two recognizable varieties occur in Illinois, distinguished by the following key:

1. Leaves minutely pubescent beneath_____1a. *S. glauca* var. *glauca*
1. Leaves glabrous beneath__________1b. *S. glauca* var. *leurophylla*

1a. **Smilax glauca** Walt. var. **glauca**

Smilax glauca var. *genuina* Blake, Rhodora 20:79. 1918.
Leaves minutely pubescent beneath.

HABITAT: Dry woods; edges of fields and bluffs.
RANGE: New Jersey to Nebraska, south to Florida and Texas.
ILLINOIS DISTRIBUTION: Rather common in the southern one-third of Illinois; absent elsewhere.

62. *Smilax glauca* (Catbrier). *a.* Habit, X⅙. *b.* Inflorescence, X½.

1b. Smilax glauca Walt. var. **leurophylla** Blake, Rhodora 20:80. 1918.

Leaves glabrous beneath.

HABITAT: Edges of swamps (in Illinois).
RANGE: Massachusetts to Missouri south to Oklahoma and Florida.
ILLINOIS DISTRIBUTION: Known from Union, Pope, and Johnson counties in extreme southern Illinois.
This more northern variety differs only in the lack of minute pubescence on the lower surface of the leaves.

2. Smilax bona-nox L. Sp. Pl. 1030. 1753.

High-climbing shrub from thick rhizomes and slender stolons; stems quadrangular, flexuous, with firm prickles or with prickles absent; leaves deltoid to ovate to panduriform, to 8 cm long, to 6 cm broad, coriaceous, never glaucous beneath, the margins thickened and sometimes spinulose; peduncles rather slender, usually ascending, to 3 cm long, exceeding the subtending petioles; berries black, glaucous, 6–8 mm in diameter, 1-seeded, the seed 4–5 mm long.

Two varieties, separated on the basis of leaf morphology, occur in Illinois.

1. Leaves spinulose, with patches of white coloration ____________ 2a. *S. bona-nox* var. *bona-nox*
1. Leaves entire or with very weak marginal spinules, green throughout ____________ 2b. *S. bona-nox* var. *hederaefolia*

2a. Smilax bona-nox L. var. **bona-nox** *Fig.* 63.

Leaves spinulose, with patches of white coloration, some of the leaves strongly panduriform.

HABITAT: Dry woods and fields; on bluffs.
RANGE: Maryland to Kansas, south to Texas and Florida; Mexico.
ILLINOIS DISTRIBUTION: Restricted to the Shawneetown Ridge of southern Illinois. More common than var. *hederaefolia*.
The strongly panduriform leaves which usually are present on each specimen are distinctive for this taxon.

63. *Smilax bona-nox* (Catbrier).—var. *bona-nox*. *a.* Habit, X¼. *b.* Flower, X1.

2b. Smilax bona-nox L. var. **hederaefolia** (Beyrich) Fern. Rhodora 46:36. 1944. *Fig. 63a.*

Smilax hederaefolia Beyrich ex Kunth, Enum. Pl. 5:209. 1850.

Leaves entire or weakly spinulose, green, some of the leaves at most only weakly panduriform.

HABITAT: Generally more moist situations than var. *bona-nox.*

RANGE: Massachusetts; Delaware to Kansas, south to Texas and Florida.

ILLINOIS DISTRIBUTION: In most localities as var. *bona-nox,* but not as plentiful.

3. Smilax rotundifolia L. Sp. Pl. 1030. 1753. *Fig. 64.*

Smilax quadrangularis Muhl. ex Willd. Sp. Pl. 4:775. 1806.

Smilax rotundifolia var. *quadrangularis* (Muhl.) Wood, Class-Book Bot. 544. 1847.

High-climbing shrub from slender rhizomes; stems terete or quadrangular, flexuous, with stout prickles, or with prickles absent; leaves ovate to suborbicular, obtuse to acute to cuspidate at apex, rounded or cordate at base, subcoriaceous, lustrous green on both surfaces, glabrous or scaberulous on the veins and margins, entire, to 12 cm long, nearly as broad, 5- to 7-nerved; peduncles rather stout, ascending or spreading, to 1.5 cm long, not much exceeding the subtending petioles; berries black, glaucous, 5–7 mm in diameter, 2- to 3-seeded.

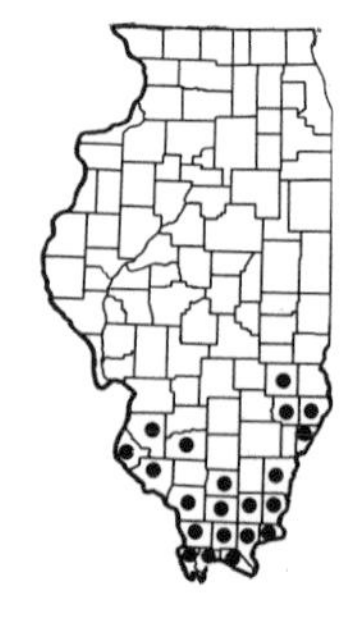

COMMON NAME: Catbrier.

HABITAT: Dry woods; edges of fields.

RANGE: Nova Scotia to Michigan, south to Florida and Texas.

ILLINOIS DISTRIBUTION: Rather common in the southern one-third of the state; absent elsewhere.

This species may be distinguished from the other woody species by the combination of flexuous stems usually with stout spines, unlobed, entire leaves green on both sides, and peduncles about the same length

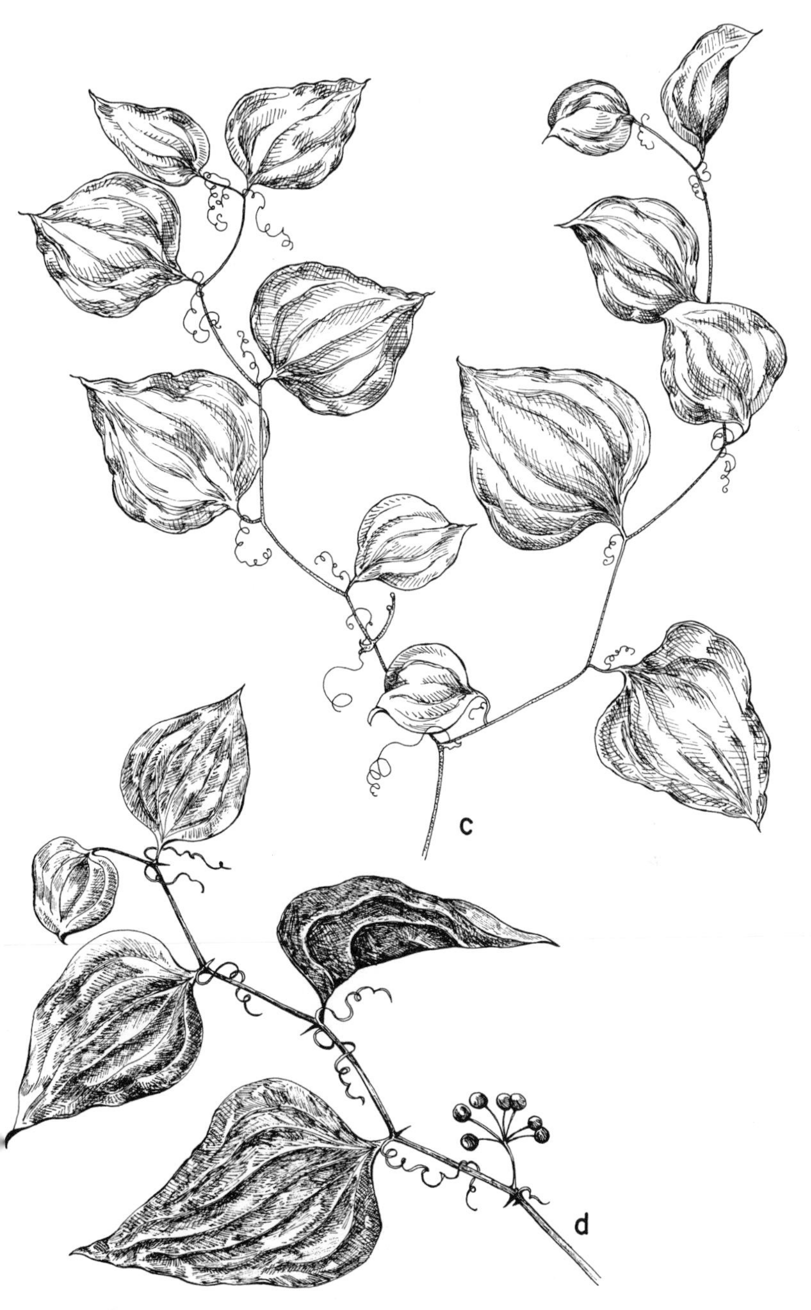

63*a*. *Smilax bona-nox* (Catbrier).—var. *hederaefolia*. *c*. Habit (sterile), X¼. *d*. Habit (in fruit), X¼.

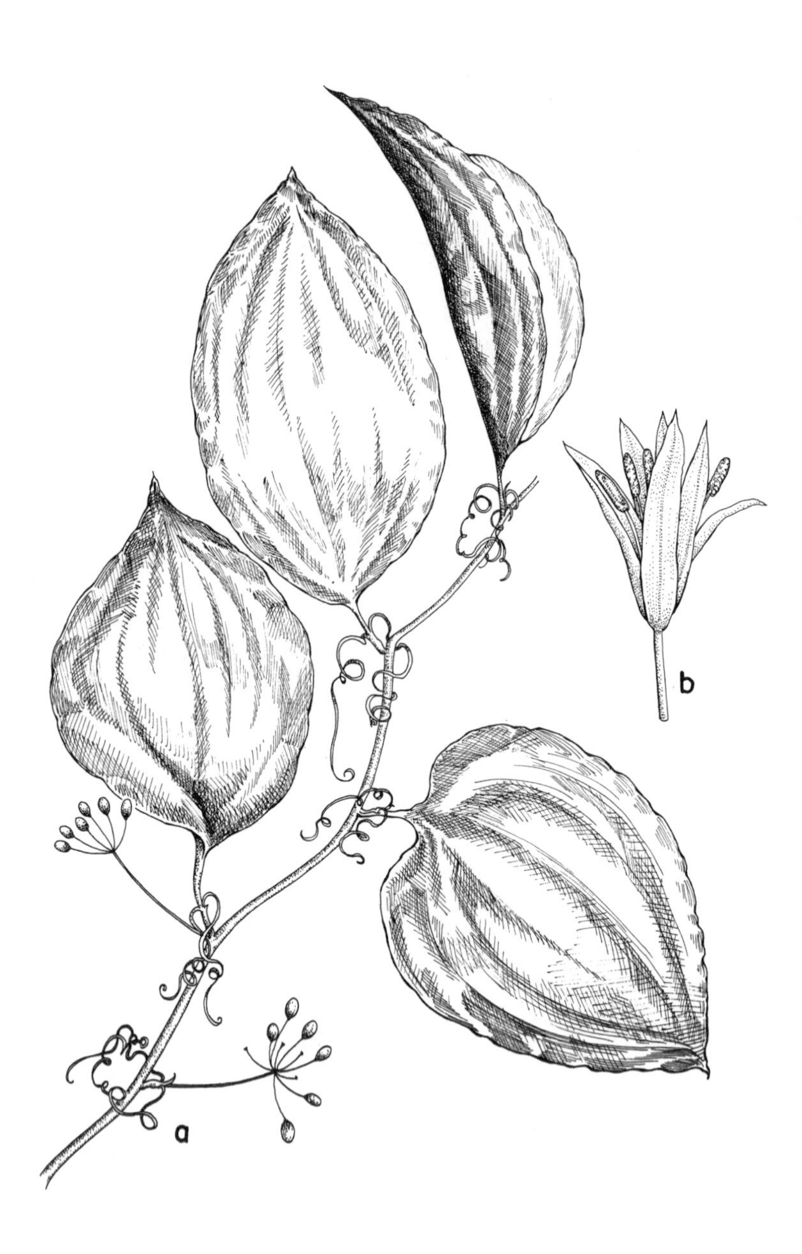

64. *Smilax rotundifolia* (Catbrier). *a.* Habit, X¼. *b.* Flower, X1.

65. *Smilax hispida* (Bristly Catbrier). *a.* Habit, X¼. *b.* Flower, X1.

as the subtending petioles. Spines are frequently absent, however.

Specimens with 4-angled stems at one time were segregated as var. *quadrangularis*, but this variation scarcely seems worthy of recognition.

4. **Smilax hispida** Muhl. ex Torr. Fl. N. Y. 2:302. 1843. *Fig. 65.*

Smilax tamnoides L. var. *hispida* (Muhl.) Fern. Rhodora 46:39. 1944.

High-climbing shrub from short, thick rhizomes; stems terete, sparsely or densely covered by short, flexible prickles; leaves orbicular to ovate to elliptical, obtuse to subacute at apex, rounded to cordate at base, to 14 (–20) cm long, to 12 (–18) cm broad, membranous, lustrous green on both sides, minutely serrulate, 5- to 7-nerved; peduncles slender, pendent, to 6 cm long, greatly exceeding the subtending petioles; berries black, 4.5–8.0 mm in diameter, 1- to 2-seeded, the seed 3–6 mm long.

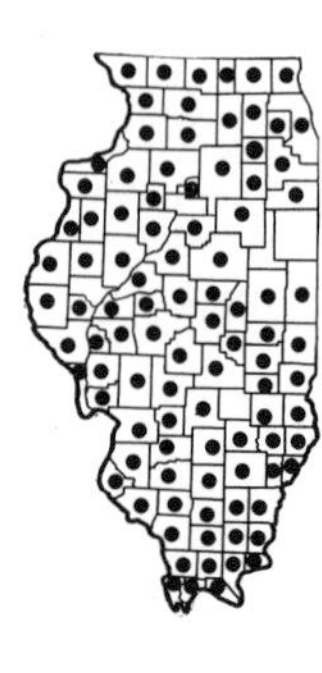

COMMON NAME: Bristly Catbrier.

HABITAT: Usually moist woods.

RANGE: Ontario to South Dakota, south to Texas and Georgia.

ILLINOIS DISTRIBUTION: Throughout the state; probably in every county.

This is the most widely distributed woody *Smilax* in Illinois.

The occasional occurrence of suborbicular leaves may serve to cause confusion between this species and *S. rotundifolia*, but *S. rotundifolia* has flexuous stems and shorter peduncles. In addition, *S. hispida* usually has blackish, weaker and shorter, prickles, often densely covering the stem.

Some authors prefer to treat this taxon merely as a variety of *S. tamnoides.*

5. **Smilax herbacea** L. Sp. Pl. 1030. 1753. *Fig. 66.*

Climbing herb with numerous tendrils; leaves narrowly ovate to orbicular, acuminate to acute to obtuse at the apex, cordate to rounded at the base, to 5 cm long, to 3.5 cm broad, pale and occasionally glaucous beneath, glabrous on both surfaces, with petioles to 5 cm long; lowest bracts bladeless, appressed-ascend-

66. *Smilax herbacea* (Carrion Flower). *a.* Habit, X¼. *b.* Fruiting cluster, X3/16.

ing; inflorescence 20- to 120-flowered; peduncle wide-spreading, more than four times as long as the subtending petioles; perianth segments 3.5 to 4.5 mm long; berries dark blue, glaucous, 7–10 mm in diameter, 3- to 6-seeded, the seeds brown; 2n = 26 (Mangaly, 1968).

COMMON NAME: Carrion Flower.

HABITAT: Near top of slope in moist woodland (in Illinois).

RANGE: Southeastern Canada to New York, southwest to Ohio, southern Illinois, western Tennessee, eastern Georgia and South Carolina.

ILLINOIS DISTRIBUTION: Known only from Jackson County (moist woods, Lake Murphysboro State Park, May 1, 1960, *R. Mohlenbrock 13317*).

Because of the glaucous berries, the glabrous leaves, and the appressed bladeless bracts at the base of the stem, it seems justifiable to maintain S. *herbacea* as a distinct species from S. *pulverulenta* and S. *lasioneunon.*

The Illinois collection of this species is nearly 200 miles west and 100 miles north of the nearest station.

6. Smilax lasioneuron Hook. Fl. Bor. Am. 2:173. 1839. *Fig. 67.*

Smilax herbacea var. *lasioneuron* (Hook.) A. DC. Monog. Phan. 1:52. 1878.

Climbing herb with numerous tendrils; leaves narrowly ovate to orbicular, obtuse and occasionally cuspidate at the apex, rounded to cordate at the base, to 7 cm long, to 5 cm broad, slightly paler on the lower surface, puberulent on the veins beneath, with petioles to 10 cm long; lowest bracts bladeless, spreading-ascending; inflorescence 10- to 100-flowered; peduncle spreading, about two times longer than the subtending petioles; perianth segments 3–5 mm long; berries dark blue, glaucous, 5–6 mm in diameter, with 3–5 seeds; 2n = 26 (Mangaly, 1968).

67. *Smilax lasioneuron* (Carrion Flower). *a.* Habit, X¼. *b.* Flower, X1.

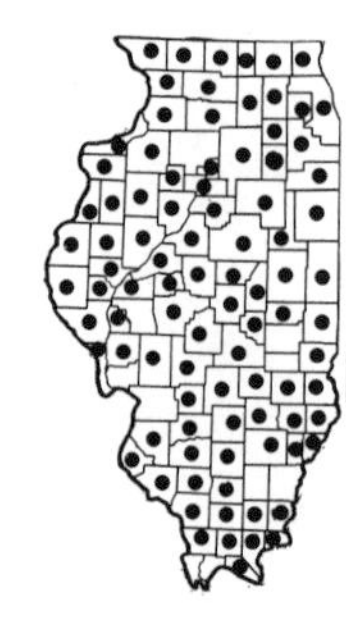

COMMON NAME: Carrion Flower.
HABITAT: Moist woods; edge of fields.
RANGE: Ontario to Saskatchewan, south to Colorado, Oklahoma, and northern Florida.
ILLINOIS DISTRIBUTION: Rather common; probably in every county in Illinois.

The pale lower leaf surface and the glaucous berries relate this species to S. *herbacea,* but the veins on the underside of the leaves of S. *lasioneuron* are puberulent. *Smilax pulverulenta* is a similar species but differs in its all green leaves, its long peduncles, and its black berries which are not glaucous.

7. **Smilax pulverulenta** Michx. Fl. Bor. Am. 2:238. 1803. *Fig.* 68.

Smilax herbacea var. *pulverulenta* (Michx.) Gray, Man. Bot. 486. 1848.

Climbing herb with numerous tendrils; leaves ovate, short-acuminate at the apex, rounded at the base, to 10 cm long, to 5 cm broad, green on both sides, puberulent on the veins beneath, with petioles to 10 cm long; lowest bracts blade-bearing; inflorescence 10- to 35-flowered; peduncle arching, at least five times longer than the subtending petioles; perianth segments 3.5–5.0 mm long; berries black, not glaucous, 8–10 mm in diameter, with 3–5 seeds; 2n = 26 (Mangaly, 1968).

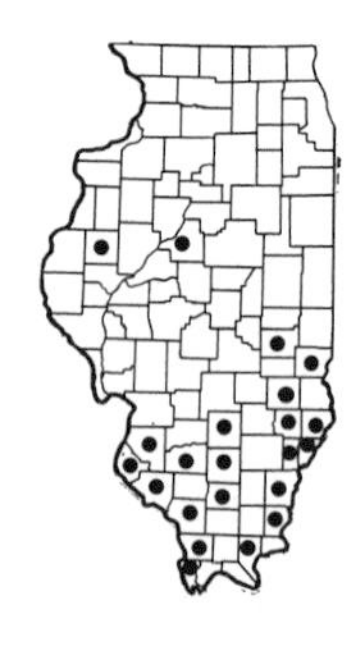

COMMON NAME: Carrion Flower.
HABITAT: Moist or dry woods; edges of fields.
RANGE: New York to Minnesota, south to Kansas, Arkansas, and South Carolina.
ILLINOIS DISTRIBUTION: Occasional in the southern two-fifths of the state; rare elsewhere.

Smilax pulverulenta appears to come into flower slightly earlier than either S. *herbacea* or S. *lasioneuron.* It has been observed in flower as early as the last of April. The extremely long peduncles are distinctive for this species.

8. **Smilax illinoensis** Mangaly, Rhodora 70:263. 1968. *Fig.* 69.

More or less upright, non-climbing herb to 1 m tall, with a few tendrils from the upper leaves; leaves variously ovate to ellipti-

68. *Smilax pulverulenta* (Carrion Flower). *a*. Habit, X¼. *b*. Inflorescence, X¼.

69. *Smilax illinoensis* (Carrion Flower). *a*. Leaves and inflorescence, X¼. *b*. Fruiting cluster, X½.

cal, acute to acuminate at the apex, truncate to subcordate at the base, to 15 cm long, to 8 cm broad, paler and pubescent beneath, with petioles to 18 cm long, always longer than the blade; lowest bracts bladeless, ascending; inflorescence 10- to 50-flowered; peduncles 5–10, slender, ascending, to 15 cm long; perianth segments 3.5–4.5 mm long; berries purple-black, to 6 mm in diameter, with 3–5 seeds; 2n = 26 (Mangaly, 1968).

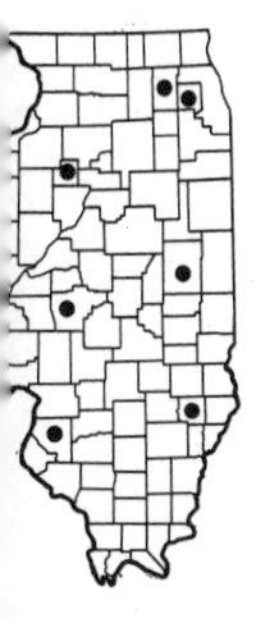

COMMON NAME: Carrion Flower.

HABITAT: Thickets often near roads.

RANGE: Ontario to Minnesota, south to central Missouri, south-central Illinois, and southwestern Ohio.

ILLINOIS DISTRIBUTION: Occasional in the northern three-fourths of the state.

This newly described species seemingly is somewhat intermediate between S. *ecirrata* and S. *lasioneuron*, although it strongly resembles the former in its erect habit. Mangaly (1968) suggests a possible hybrid origin for S. *illinoensis*.

9. **Smilax ecirrata** (Engelm.) S. Wats. in Gray, Man. 520. 1890. *Fig. 70.*

Coprosmanthus herbaceus β ecirratus Engelm. ex Kunth, Enum. 5:266. 1850.

Smilax herbacea var. *ecirrata* (Engelm. ex Kunth) A. DC. Monogr. Phan. 1:52. 1878.

More or less upright, non-climbing herb to nearly 1 m tall, with a few tendrils from the upper leaves only or with tendrils completely lacking; leaves broadly ovate to suborbicular, acute to obtuse and occasionally cuspidate at the apex, rounded to cordate at the base, to 15 cm long, to 10 cm broad, paler and pilosulous beneath (at least when young), with petioles to 10 cm long but always shorter than the blade; lowest bracts bladeless, ascending; inflorescence 6- to 20-flowered; peduncles 1–3, very slender, ascending, to 10 cm long; perianth segments 3.5–4.0 mm long; berries purple-black, to 6 mm in diameter, with 3–5 seeds; 2n = 26 (Mangaly, 1968).

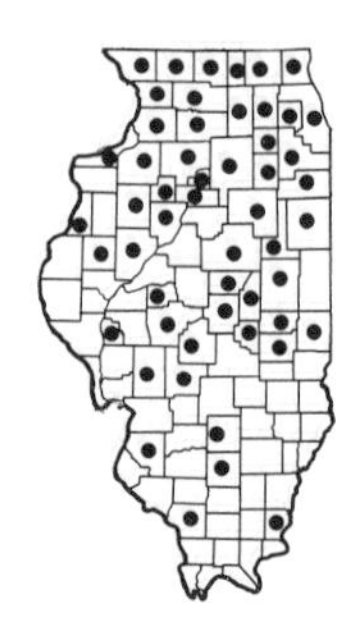

COMMON NAME: Carrion Flower.

HABITAT: Moist woods.

RANGE: Ontario to Minnesota, south to northeastern Oklahoma and Tennessee.

ILLINOIS DISTRIBUTION: Occasional in the northern one-half of Illinois, becoming extremely uncommon in the southern counties.

Smilax ecirrata, together with *S. illinoensis,* are the only non-climbing species of the genus in Illinois. Tendrils often are lacking. The inflorescence contains fewer flowers than in any of the other herbaceous species in Illinois. *Smilax illinoensis* differs by its proportionately longer petioles and its narrower leaves.

The flowers appear during May and the first part of June.

Although the epithet is customarily spelled *ecirrhata,* the original spelling was without the "h". The lectotype, selected by Mangaly, was collected by G. Engelmann in 1835 from Belleville, Illinois.

DIOSCOREACEÆ—WILD YAM FAMILY

Only the following genus occurs in Illinois.

1. *Dioscorea* L. – Wild Yam

Twining herbaceous vines from rhizomes and tubers; leaves alternate or whorled, net-veined; plants dioecious; inflorescence axillary, the staminate glomerulate or solitary in panicles, the pistillate solitary in interrupted spikes; flowers unisexual, regular; perianth parts 6, of uniform color; stamens 3 or 6; ovary 3-celled, inferior, with 2 ovules per cell; capsule 3-winged, loculicidal; seeds flat, broadly winged.

The cultivated cinnamon vine, *D. batatas* Dcne., is grown as an ornamental in Illinois, but has not been collected as an escape.

KEY TO THE SPECIES OF Dioscorea IN ILLINOIS

1. All leaves (except sometimes the lowermost) alternate; capsule 1.5–2.5 cm long; seeds (including wing) 7–14 mm broad; petiole essentially glabrous at point of attachment of blade___1. *D. villosa*
1. Lowest leaves whorled, becoming opposite or alternate on the upper part of the stem; capsule 2.5–3.0 cm long; seeds (including wing) 15–18 mm broad; petiole puberulent at point of attachment of blade____2. *D. quaternata*

70. *Smilax ecirrata* (Carrion Flower). *a.* Habit, X¼. *b.* Inflorescence, X¼. *c.* Flower, X1¼.

1. **Dioscorea villosa** L. Sp. Pl. 1033. 1753. *Fig. 71.*

Dioscorea paniculata Michx. Fl. Bor. Am. 2:239. 1803.

Dioscorea paniculata var. *glabrifolia* Bartlett, Bull. U. S. Bur. Pl. Ind. 189:15. 1910.

Dioscorea villosa f. *glabrifolia* (Bartlett) Fern. Rhodora 39: 401. 1937.

Leaves alternate, the lowest becoming pseudo-whorled, ovate, acuminate, cordate, 5–10 cm long, 3–8 cm broad, 7- to 11-nerved, glabrous above, glabrous or puberulent beneath; petiole to 7 cm long, glabrous or nearly so; staminate inflorescence paniculate, the internodes between the glomerules 2–4 mm long; pistillate inflorescence spicate, interrupted, 5- to 15-flowered; capsule 1.5–2.5 cm long, 3-winged; seeds broadly winged, 7–14 mm broad (including the wing); 2n = 60 (Smith, 1937).

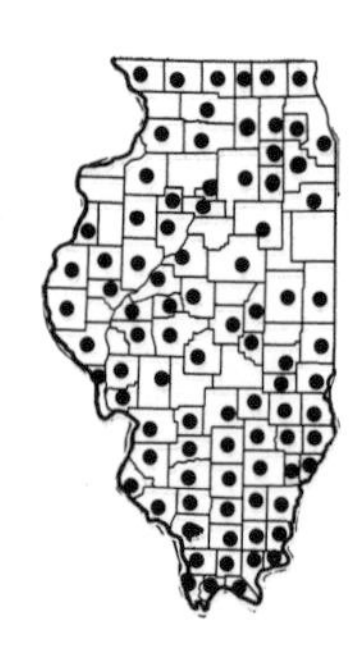

COMMON NAME: Wild Yam.

HABITAT: Dry or moist woodlands.

RANGE: Connecticut to Minnesota, south to Texas and Florida.

ILLINOIS DISTRIBUTION: Common; probably in every county, although so far not collected in every county.

This wild yam has large, thickened rhizomes which are rarely seen because of the great depth where they grow in the soil. Specimens with completely glabrous leaves may be known as f. *glabrifolia.* A few specimens have been observed, however, where both glabrous and puberulent leaves occur on the same plant. No attempt has been made to plot the distribution of Illinois material based on leaf pubescence. Flowers appear in June and July, with fruits maturing by late August.

2. **Dioscorea quaternata** [Walt.] J. F. Gmel. Syst. 581.1796. *Fig. 72.*

Anonymos quaternata Walt. Fl. Car. 246. 1788.

Dioscorea glauca Muhl. Cat. 92. 1813.

Dioscorea quaternata var. *glauca* (Muhl.) Fern. Rhodora 39:399. 1937.

Lowest leaves in whorls of 4–7, the middle and upper becoming opposite or alternate, ovate, acuminate, cordate, 5–12 cm long, 4–10 cm broad, 7- to 11-nerved, green or scarcely glaucous, glabrous above, glabrous or puberulent beneath; petiole 3–9 cm long, puberulent at least at point of blade attachment; staminate

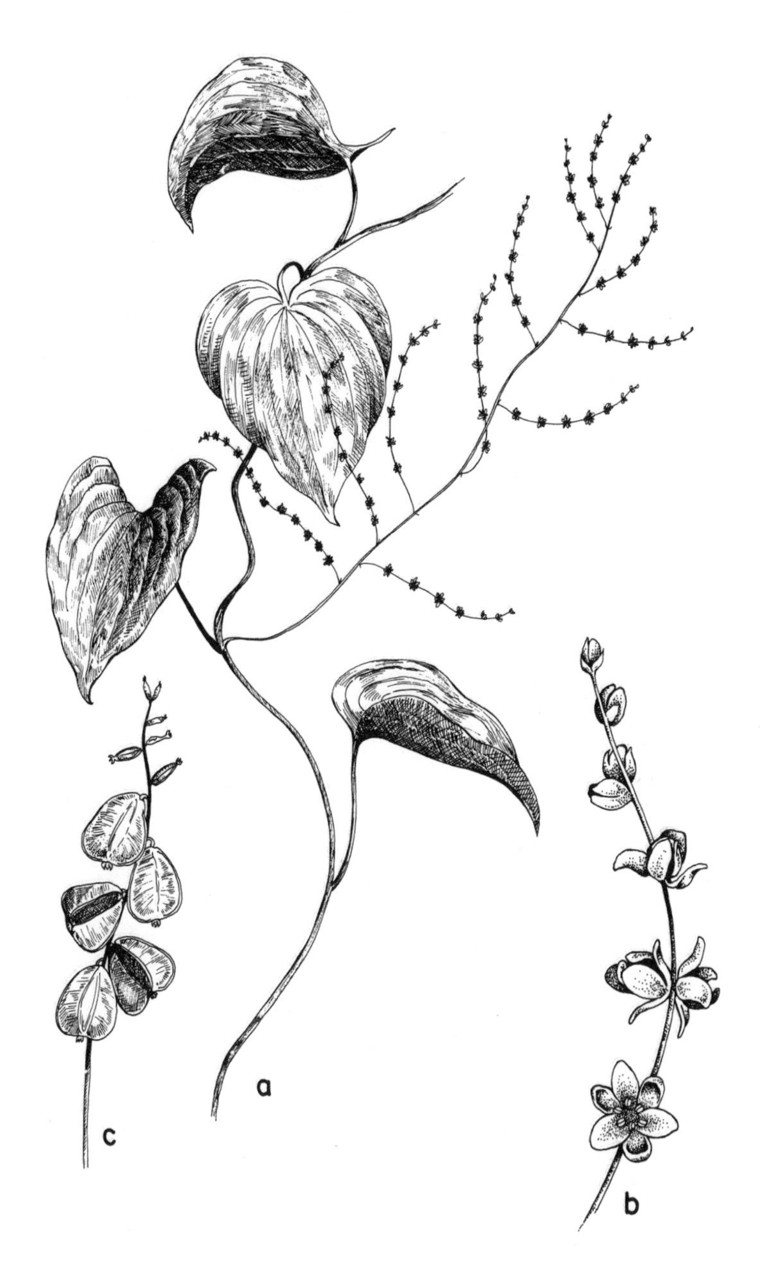

71. *Dioscorea villosa* (Wild Yam). *a.* Habit, X⅙. *b.* Staminate inflorescence, X2½. *c.* Pistillate inflorescence and young fruits, X2½.

inflorescence paniculate, the internodes between the glomerules 3–10 mm long; pistillate inflorescence spicate, interrupted, 3- to 10-flowered; capsule 2.5–3.0 cm long, 3-winged; seeds broadly winged, 15–18 mm broad (including the wing).

COMMON NAME: Wild Yam.

HABITAT: Dry or moist woodlands.

RANGE: Pennsylvania to Missouri, south to Oklahoma and Florida.

ILLINOIS DISTRIBUTION: Occasional; confined to the southern one-third of the state.

Plants with the leaves vaguely glaucous have been called var. *glauca.* This character, which at best is only poorly developed, is virtually impossible to detect with herbarium material. It is not distinguished in this work.

IRIDACEÆ—IRIS FAMILY

Perennial herbs with rhizomes or fibrous roots; flowers regular, perfect; perianth parts 6, united at least below, generally uniform in color; stamens 3; ovary inferior; fruit a capsule, 3-celled, loculicidal.

The Iridaceae differs from the showy-flowered Liliaceae in the reduction of its stamens to three. The leaves frequently are folded lengthwise and generally are sword-shaped (ensiform).

Besides the native or escaped species enumerated below, several species of *Iris, Gladiolus, Crocus, Freesia,* and *Tigridia* are grown as ornamentals in Illinois.

KEY TO THE GENERA OF Iridaceae IN ILLINOIS

1. Sepals somewhat recurved; petals spreading or erect; styles petal-like; stamens concealed by the arching styles............1. *Iris*
1. Sepals and petals similar, spreading or tubular; styles not petal-like; stamens not concealed.
   2. Perianth tubular, forming a short funnel, scarlet; upper three perianth segments longer than the lower three....2. *Gladiolus*
   2. Perianth spreading and orange, or rotate and blue or white; all perianth segments essentially equal.
      3. Flowers orange, with purplish spots, at least 3 cm broad; filaments distinct; flowers bracteate, the bracts soon withering; capsule at least 2 cm long; seeds attached to a central column, resembling a blackberry; plants rhizomatous..3. *Belamcanda*

72. *Dioscorea quaternata* (Wild Yam). *a.* Habit, X⅙. *b.* Staminate inflorescence, X2½. *c.* Capsule, X½.

3. Flowers blue or white, less than 2 cm broad; filaments united to apex; flowers borne from 1 or 2 persistent spathes; capsule up to 6 mm long; seeds not attached to a central column; plants with fibrous roots________________4. *Sisyrinchium*

1. *Iris* L. – Iris

Perennial herbs from usually stout rhizomes; leaves ensiform, usually folded lengthwise, not confined to the base of the plant; inflorescence 1- to several-flowered, the flowers arising from spathes; perianth parts 6, of uniform color, clawed, the outer sepals generally slightly larger, more or less recurved, the inner petals spreading or erect; stamens 3, attached to the base of the sepals, concealed by the styles; styles petal-like, arching, bilobed; stigma flat, located between the lobes of the style; ovary 3- to 6-angled; capsule 3- to 6-lobed, with 1–2 seeds per cell.

Of the many varieties and species cultivated as garden ornamentals, probably *I. germanica,* the bearded iris, is most common. Several species of cultivated *Iris* have the ability to develop rapidly and become established in our flora.

KEY TO THE SPECIES OF Iris IN ILLINOIS

1. Upper surface of sepals with a beard of hairs.
   2. Stems very short or none, bearing one flower; leaves to 15 cm long, to 7 mm broad; sepals to 17 mm broad______1. *I. pumila*
   2. Stems to nearly 1 m tall, bearing several flowers; leaves to 100 cm long, to 3 cm broad; sepals over 20 mm broad____2. *I. germanica*
1. Sepals without a beard of hairs.
   3. Rhizomes stout, at least 1 cm thick; leaves 40–100 cm long; flowering stem usually branched, 20–100 cm long; spathes subtending flower unequal in length; flowers several per stem; sepals 1- to 2-ridged, or without ridges; capsule obtusely 3-angled or 6-angled, 3–9 cm long; plants often of more aquatic situations.
      4. Ovary and capsule 6-angled; capsule indehiscent; leaves rather soft.
         5. Flowering stem 20–40 cm tall; flowers dark blue; capsule 3–5 cm long; leaves 15–30 mm broad; seeds more or less globoid______________________________3. *I. brevicaulis*
         5. Flowering stem at least 50 cm tall; flowers coppery; capsule 4.5–7.5 cm long; leaves 10–15 mm broad; seeds flattened__________________________________4. *I. fulva*

4. Ovary and capsule 3-angled (occasionally 6-angled in the yellow-flowered *I. pseudacorus*); capsule dehiscent; leaves firm.
    6. Flowers usually some shade of blue or violet; sepals not 2-ridged on upper surface; perianth tube constricted above the ovary________________________________5. *I. shrevei*
    6. Flowers basically yellow; sepals 2-ridged on upper surface; perianth not constricted above the ovary______________ __________________________________6. *I. pseudacorus*
3. Rhizomes slender, less than 1 cm thick; leaves (at flowering time) to 25 cm long, at maturity as long as 40 cm; flowering stem unbranched, to 7 cm long; spathes subtending flower nearly equal in length; flowers 1–2 per stem; sepals sharply 3-ridged above; capsule sharply 3-angled, about 1 cm long; plants of rich woodlands, usually near streams____7. *I. cristata*

1. **Iris pumila** L. Sp. Pl. 38. 1753. *Fig. 73.*

Plants forming dense clumps; leaves to 15 cm long, to 7 mm broad; stem very short or none, bearing a single flower; flowers not fragrant, typically reddish-purple with bluish or yellowish beard; sepals to 17 mm broad, recurved, densely bearded; petals ascending, about as long as the sepals; capsule not observed.

COMMON NAME: Dwarf Iris.

HABITAT: Escaped from cultivation along railroads and other waste land.

RANGE: Central Europe to Asia Minor; occasionally escaped from cultivation in the United States.

ILLINOIS DISTRIBUTION: Known from several collections by Umbach from DuPage County between 1898 and 1903.

This is an extremely low-growing, rapidly spreading iris with very short, slender leaves. Aside from the robust *Iris germanica, Iris pumila* is the only *Iris* escaped in Illinois with bearded sepals.

2. **Iris germanica** L. Sp. Pl. 38. 1753. *Fig. 74.*

Rhizome stout; leaves up to 100 cm long, to 3 cm broad, glaucous; flowering stem to nearly 1 m tall, mostly branched, several-flowered; spathes unequal in length; flowers 7–10 cm broad, usually some shade of violet marked with yellow, white, and

73. *Iris pumila* (Dwarf Iris). X¼.

74. *Iris germanica* (Blue Flag). X¼.

brown veins; sepals broadly ovate, recurved, with a dense beard or hairs along the central nerve on the upper surface; petals ascending, slightly smaller than the sepals; capsule 3-angled.

COMMON NAME: Bearded Iris; Blue Flag.
HABITAT: Escaped from cultivation into waste land.
RANGE: Original area unknown; now only found in cultivation or as an escape.
ILLINOIS DISTRIBUTION: Occasionally found throughout the state, but infrequently collected.
This is a common cultivated species of Iris. Many color variations are grown. It is the only iris with a beard of hairs on the sepals.

3. **Iris brevicaulis** Raf. Fl. Ludov: 20. 1817. *Fig. 75.*

Iris foliosa Mack. & Bush, Trans. Acad. Sci. St. Louis 12:81. 1902.

Rhizome very stout; leaves 40–70 cm long, 15–30 mm broad, green; flowering stem 20–40 cm long, simple or branched, zigzag, several-flowered; spathes unequal in length; flowers 8–10 cm broad, deep blue; sepals ovate, papillose, marked with yellow and white, recurved or spreading; petals oblanceolate, spreading; ovary 6-angled; capsule 6-angled, 3–5 cm long, indehiscent.

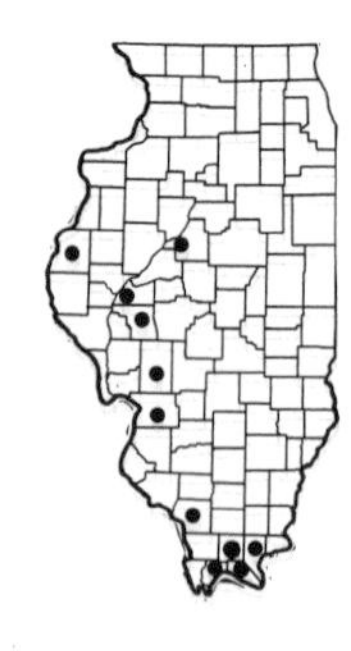

COMMON NAME: Blue Iris.
HABITAT: Marshes and wet prairies.
RANGE: Ohio to Kansas, south to Texas and Alabama.
ILLINOIS DISTRIBUTION: Not common; scattered throughout the state, except in northern one-third of Illinois.
This is the largest-flowered iris in Illinois, with the flowers blooming from late May to mid-July. The flowering stems are surpassed by the leaves so that the flowers sometimes appear to be hidden. Little variability has been observed in this species in Illinois.

4. **Iris fulva** Ker, Bot. Mag. 36:pl. 1496. 1812. *Fig. 76.*

Iris cuprea Pursh, Fl. Am. Sept. 30. 1814.

Rhizome stout; leaves 40–80 cm long, 10–15 mm broad, green;

75. *Iris brevicaulis* (Blue Iris). *a.* Inflorescence and leaves, X⅙. *b.* Capsule, X¼.

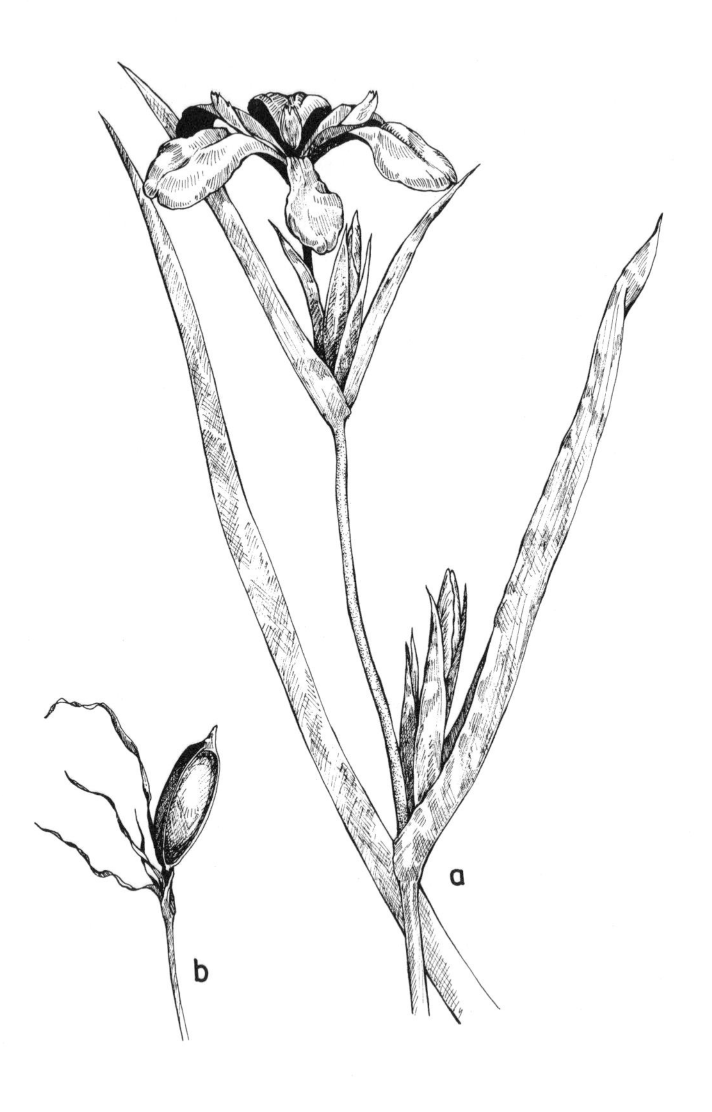

76. *Iris fulva* (Swamp Iris). *a.* Inflorescence and leaves, X¼. *b.* Capsule, X⅕.

flowering stem at least 50 cm tall, simple or branched, several-flowered; spathes unequal in length; flowers 7–9 cm broad, copper; sepals obovate, recurved; petals obovate, slightly smaller than the sepals, spreading; ovary 6-angled; capsule 6-angled, 4.5–7.5 cm long, indehiscent.

COMMON NAME: Swamp Iris; Red Iris.
HABITAT: Swamps, usually in shallow water.
RANGE: Illinois and Missouri, south to Louisiana and Georgia.
ILLINOIS DISTRIBUTION: Rare; limited to five counties in extreme southern Illinois.

The swamp iris is a species of the Gulf Coast region which has made its way north as far as southeastern Missouri and southern Illinois. The copper-colored flowers, produced from mid-May to early June, distinguish this species. It is restricted to swampy situations.

5. **Iris shrevei** Small, Addisonia 12:13. 1927. *Fig. 77.*

Iris virginica var. *shrevei* (Small) E. Anders. Ann. Mo. Bot. Gard. 23:469. 1936.

Rhizome stout; leaves 40–100 cm long, up to 3 cm broad, green; flowering stem 60–100 cm tall, branched or unbranched, several-flowered; spathes unequal in length; flowers 6–8 cm broad, basically bluish or violet; sepals spatulate, clawed, recurved, pubescent and yellowish near the base; petals a little smaller, spreading to ascending; ovary 3-angled; capsule obtusely 3-angled, 6–9 cm long, dehiscent.

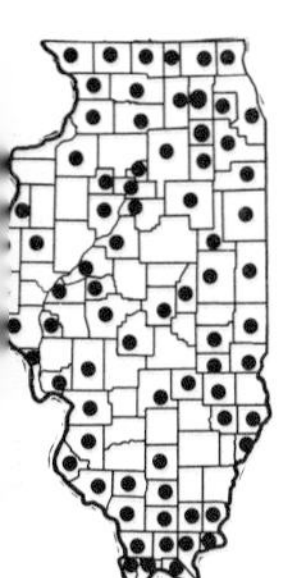

COMMON NAME: Blue Iris; Wild Flag.
HABITAT: Wet situations.
RANGE: Ontario to Minnesota, south to Texas and Alabama.
ILLINOIS DISTRIBUTION: Common; probably in every county.

This is the only common native species of *Iris* in Illinois. It flowers from early May until about mid-June. Some authors consider this iris to be a variety of the southeastern *Iris virginica.* The slightly thicker and somewhat more elongate, prominently beaked capsule is the chief distinction of *I. shrevei* from *I. virginica.*

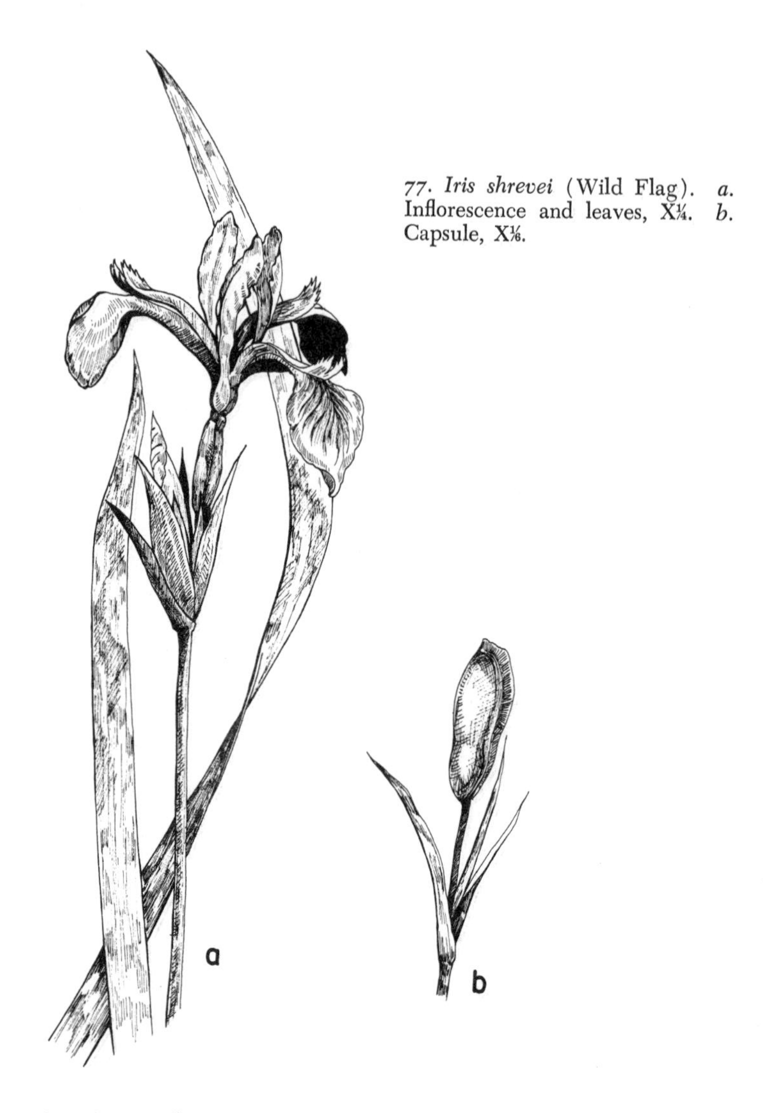

77. *Iris shrevei* (Wild Flag). *a.* Inflorescence and leaves, X¼. *b.* Capsule, X⅙.

6. Iris pseudacorus L. Sp. Pl. 38. 1753. *Fig. 78.*

Rhizome stout; leaves 50–100 cm long, about 1.5 cm broad, green; flowering stem at least 50 cm tall, branched or unbranched, several-flowered; spathes unequal in length; flowers 7–9 cm broad, basically yellow, with dark brown markings; sepals obovate, 2-ridged above, spreading; petals much smaller, narrowed at the middle, spreading or ascending; ovary usually 3-angled; capsule 3-angled or occasionally 6-angled, 5–8 cm long, dehiscent; 2n = 34 (Gadella & Kliphuis, 1963; Sorsa, 1963).

78. *Iris pseudacorus* (Yellow Iris). X¼.

COMMON NAME: Yellow Iris.
HABITAT: Escaped into waste ground and along roads.
RANGE: Native of Europe; found throughout eastern North America.
ILLINOIS DISTRIBUTION: Scattered throughout the state. This is another cultivated species which appears to be able to establish itself readily in low, weedy areas. The yellow flowers are produced from June to mid-August.

7. **Iris cristata** Ait. Hort. Kew. 1:70. 1789. *Fig. 79.*

Rhizome slender, less than 1 cm in diameter; leaves 10–20 cm long (at flowering time), becoming up to 40 cm long at maturity, 1.0–2.5 cm broad; flowering stems to 7 cm tall, simple, 1- to 2-flowered; spathes nearly equal in length; flowers 6–8 cm broad, basically lilac or purple, spotted with white and orange; sepals obovate, 3-ridged, spreading; petals oblanceolate, smaller than the sepals, spreading; ovary 3-angled; capsule sharply 3-angled, about 1 cm long, dehiscent.

COMMON NAME: Dwarf Crested Iris.
HABITAT: Lowland woods, usually along streams.
RANGE: Maryland to Missouri, south to Oklahoma and Georgia.
ILLINOIS DISTRIBUTION: Rare; known only from five counties in the extreme southern part of the state. At the Massac Tower station in Pope County, *Iris cristata* forms an extensive colony covering 25 square feet.

The small stature and the beautiful flower make this one of the most lovely wild flowers in Illinois. It was first discovered in Illinois in 1932. The flowering time is late April to mid-May, making it the first *Iris* to flower in Illinois.

2. *Gladiolus* L. – Gladiolus

Perennial herbs from tunicated corms; leaves ensiform, not confined to the base of the plant; inflorescence several-flowered, with each flower sessile and subtended by a spathe-like bract; perianth parts 6, of uniform color, funnelform, the upper 3 segments larger than the lower 3; stamens 3; styles not petal-like; ovary inferior, 3-celled; capsule 3-lobed, loculicidally dehiscent.

Only the following adventive hybrid occurs in Illinois.

79. *Iris cristata* (Dwarf Crested Iris). *a*. Habit, X¼. *b*. Fruiting plant, X⅜.

1. Gladiolus × colvillei Sweet, Brit. Flow. Gard. t. 155. *Fig. 80.*

Leaves to 75 cm long, to 2.5 cm broad; stem up to about 1 m tall, erect, unbranched; inflorescence short-spicate, bracteate; flowers widely expanding at apex, 2–4 cm broad, scarlet with yellow patches near base; sepals and petals oblong, acute; capsule dehiscent; 2n = 30 (Bamford, 1935).

COMMON NAME: Scarlet Gladiolus.
HABITAT: Escaped along moist roadsides (in Illinois).
RANGE: Not originally found in the wild.
ILLINOIS DISTRIBUTION: Escaped along roads in Johnson and Massac counties.

The scarlet gladiolus is a handsome, brilliant, relatively small-flowered hybrid *Gladiolus*. It is a hybrid between two south African species, *G. concolor* Salisb. and *G. cardinalis* Curt., and was first formed in cultivation. The hybrid establishes itself in disturbed areas very well, and is considered an adventive member of the flora of several of the southern states.

The flowers appear in May and June.

3. *Belamcanda* ADANS. – Blackberry Lily

Perennial herbs from stout rhizomes; leaves ensiform, resembling *Iris*, not confined to the base of the plant; inflorescence 1- to several-flowered, bracteate; perianth parts 6, of uniform color, spreading; stamens 3, united only at the base; styles not petal-like; capsule 3-lobed, dehiscent; seeds attached to a central column.

Only the following species occurs in Illinois.

1. Belamcanda chinensis (L.) DC. in Redoute, Les Liliacees 3:pl. 121. 1807. *Fig. 81.*

Ixia chinensis L. Sp. Pl. 36. 1753.

Pardanthus chinensis (L.) Der. in Konig & Sims, Ann. Bot. 1:246. 1805.

Leaves 50–70 cm long, to 2.5 cm broad; stem to 60 cm tall, branched or unbranched; inflorescence cymose, several-flowered, bracteate, the bracts soon withering; flowers 3–5 cm broad, orange spotted with purple; sepals and petals broadly elliptic,

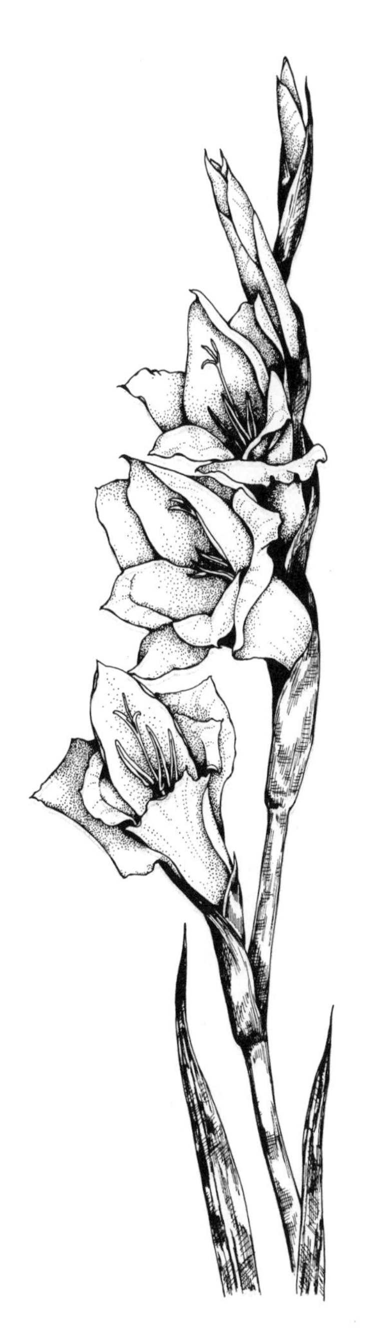

80. *Gladiolus* × *colvillei* (Scarlet Gladiolus). X⅜.

subacute; capsule 2–3 cm long, dehiscent, the valves strongly recurved after dehiscence and soon falling; seeds several, globoid, black, clustered along a central column.

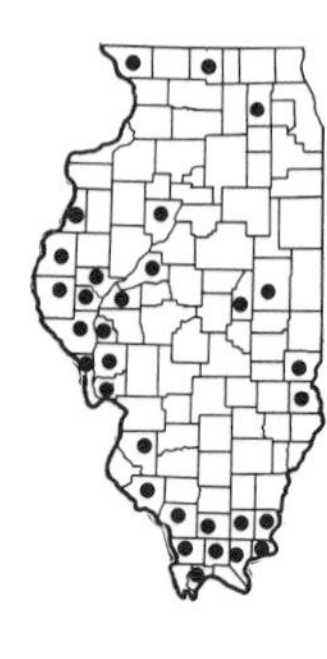

COMMON NAME: Blackberry Lily.
HABITAT: Roadsides.
RANGE: Native of Asia; frequently escaped from cultivation in the eastern United States.
ILLINOIS DISTRIBUTION: Occasional; scattered throughout the state.

The orange perianth with purple spots resembles the flower of some lilies, while the clustered black seeds are reminiscent of a blackberry; hence the name blackberry lily.

4. *Sisyrinchium* L. – Blue-eyed Grass

Perennial herbs from fibrous roots; stem often winged; leaves linear, mostly from the base of the plant; inflorescence more or less umbellate, several-flowered, produced from 1 or 2 spathes; perianth parts 6, rotate, of uniform color, withering rapidly; filaments united nearly to summit; capsule small, loculicidal.

The species of this genus do not differ greatly one from another. In many instances, only a single good difference may be found between species. The key which follows is similar to that found in most manuals simply because there are few distinguishing characters one may use in constructing the key.

The rapidly withering perianth causes this structure to be virtually useless in identification.

There is a continuation of the stem beyond the inflorescence in the form of a leaf-like bract. In two of our species, the spathes are long-pedunculate from the axil of this bract; in the other species, the spathe is sessile and terminates the stem.

Bicknell's treatment (1896) of the genus, which is too liberal in the splitting off of species, has been basically followed in this work.

KEY TO THE SPECIES OF Sisyrinchium IN ILLINOIS

1. Spathes long-pedunculate from the axils of leaf-like bracts; capsules dark brown.
   2. Leaves bright green; stems broadly winged, 3–5 mm wide; capsule 4–6 mm long; plants becoming black on drying---------- ---------------------------------------1. *S. angustifolium*

81. *Belamcanda chinensis* (Blackberry Lily). *a.* Inflorescence, X¼. *b.* Leaves, X¼. *c.* Mature capsule with seeds, X¼.

2. Leaves pale green or glaucous; stems narrowly winged, 1–3 mm wide; capsule 3.0–4.5 mm long; plants generally not becoming black on drying________________________________2. *S. atlanticum*

1. Spathes sessile, terminal; capsules pale brown, stramineous, or greenish (occasionally becoming dark brown in *S. montanum*).
   3. Spathes 2____________________________________3. *S. albidum*
   3. Spathe 1.
      4. Some or all the stems less than 3 mm wide; capsules 2–4 mm long; margins of outer bract free to base or united for less than 2 mm.
         5. Plants pale green or glaucous, drying pale; stems winged; margins of outer bract free to base; flowers pale blue or white________________________________4. *S. campestre*
         5. Plants dark green, drying blackish; stems marginate; margins of outer bract united for 1–2 mm above base; flowers bright violet________________5. *S. mucronatum*
      4. Stems 3–4 mm wide; capsules 4–6 mm long; margins of outer bract united for 2–6 mm above base_______6. *S. montanum*

1. **Sisyrinchium angustifolium** Mill. Gard. Dict. 8:2. 1768. *Fig. 82.*

Sisyrinchium graminoides Bickn. Bull. Torrey Club 23:133. 1896.

Plants bright green, drying blackish; leaves 1.5–5.0 mm broad; stems to 50 cm tall, broadly winged, 3–5 mm wide; spathes long-pedunculate from the axils of leaf-like bracts; perianth pale blue or violet-blue; capsule subgloboid, 4–6 mm long, dark brown.

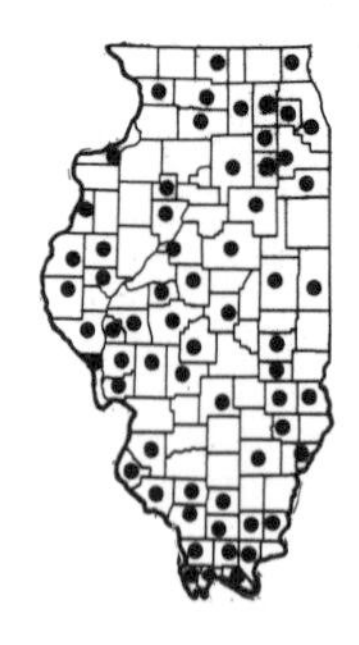

COMMON NAME: Blue-eyed Grass.

HABITAT: Low woodlands; moist prairies.

RANGE: Newfoundland to Ontario, south to Texas and Florida.

ILLINOIS DISTRIBUTION: Common, in nearly every county of the state.

Although this species has been called *S. graminoides* or *S. bermudiana* or *S. angustifolium*, Ward (1968) has presented evidence why *S. angustifolium* is the correct binomial.

The flowers are produced in May and June, occasionally persisting until mid-July. The long-pedunculate spathes and the broadly winged stems are the distinguishing features of this species.

82. *Sisyrinchium angustifolium* (Blue-eyed Grass). *a.* Inflorescence and leaves, X¼. *b.* Capsule, X2½. *c.* Seed, X10.

2. **Sisyrinchium atlanticum** Bickn. Bull. Torrey Club 23:134. 1896. *Fig. 83.*

Sisyrinchium apiculatum Bickn. Bull. Torrey Club 26:300. 1899.

Plants pale green or glaucous, drying pale; leaves 1–3 mm broad; stems to 70 cm tall, marginate, 1–3 mm wide; spathes long-pedunculate from the axils of leaf-like bracts; perianth violet-blue; capsule subgloboid, 3–5 mm long, dark brown.

COMMON NAME: Blue-eyed Grass.

HABITAT: Wet prairies.

RANGE: Nova Scotia to Michigan, south to Texas and Mississippi.

ILLINOIS DISTRIBUTION: Very rare; known only from Kankakee County (*R. A. Schneider s.n.*)

The paler aspect of the plants, the narrower stems, and the slightly smaller capsules distinguish this species from *S. angustifolium*. It flowers during May and June.

3. **Sisyrinchium albidum** Raf. Atl. Journ. 17. 1832. *Fig. 84.*

Sisyrinchium bermudianum var. *albidum* (Raf.) Babcock, Lens 1:222. 1872.

Plants pale green and usually glaucous, drying pale; leaves 1–3 mm broad; stems to 45 cm tall, broadly or narrowly winged, to 4 mm wide; spathes sessile, 2; perianth whitish or pale violet; capsule subgloboid, 2–4 mm long, pale brown or stramineous.

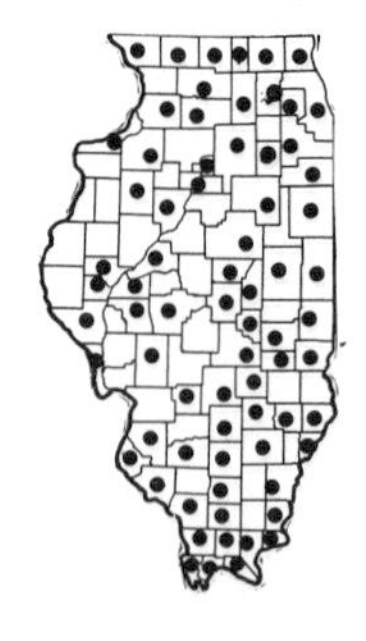

COMMON NAME: Blue-eyed Grass.

HABITAT: Open woodlands; fields; prairies.

RANGE: New York to North Dakota, south to Oklahoma and Georgia.

ILLINOIS DISTRIBUTION: Common; probably in every county except perhaps some western counties.

In the prairies of northern Illinois, the blue-eyed grass is associated with *Andropogon scoparius, Comandra richardsiana, Dodecatheon meadia, Heuchera richardsonii, Hypoxis hirsuta, Lithospermum canescens, Lobelia spicata, Pedicularis canadensis, Phlox pilosa, Silphium integrifolium* var. *deamii, Silphium terebinthinaceum, Stipa spartea*, and *Zizia aurea*.

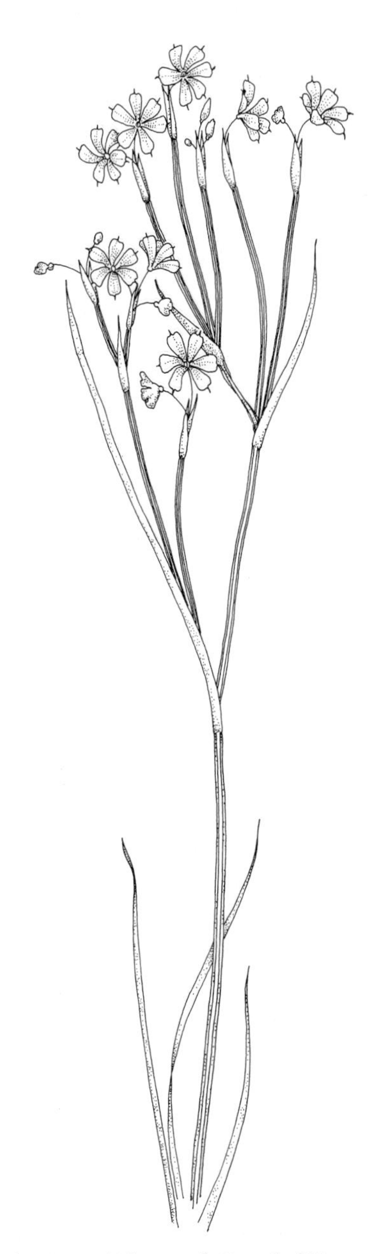

83. *Sisyrinchium atlanticum* (Blue-eyed Grass). X¼.

84. *Sisyrinchium albidum* (Blue-eyed Grass). *a.* Habit, X¼. *b.* Fruiting cluster, X2. *c.* Seed, X10.

This is the most abundant of the sessile-spathed species of *Sisyrinchium* in Illinois. It flowers from the last of April into June. It is the only species with two spathes in the sessile-spathed group.

4. **Sisyrinchium campestre** Bickn. Bull. Torrey Club 26:341. 1899. *Fig.* 85.

Plants pale green or glaucous, drying pale; leaves 1–3 mm broad; stems to 50 cm tall, winged, 1.0–2.5 mm wide; spathe sessile, 1, the outer margin free to base; perianth pale blue to whitish; capsule subgloboid, 2–4 mm long, pale brown or stramineous.

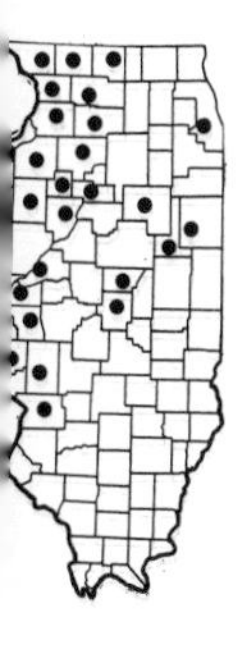

COMMON NAME: Blue-eyed Grass.
HABITAT: Prairies, particularly in sandy soil.
RANGE: Wisconsin to Minnesota, south to New Mexico and Louisiana.
ILLINOIS DISTRIBUTION: Occasional; restricted to northwestern and west-central counties. Unlike *S. mucronatum*, which turns blackish upon drying, this glaucous prairie species remains pallid when dry. The stems in *S. campestre* are more broadly winged than in *S. mucronatum*, also.

5. **Sisyrinchium mucronatum** Michx. Fl. Bor. Am. 2:33. 1803. *Fig.* 86.

Plants dark green, drying blackish; leaves 1–2 mm broad; stems to 45 cm tall, marginate, 0.5–1.5 cm wide; spathe sessile, 1, the margin united for 1–2 mm above the base; perianth violet; capsule globoid, 2–4 mm long, greenish or stramineous.

COMMON NAME: Blue-eyed Grass.
HABITAT: Sandy prairies.
RANGE: Maine to Minnesota, south to Illinois and North Carolina.
ILLINOIS DISTRIBUTION: Rare; restricted to Hancock (Augusta, *S. B. Mead*) and Mason (W of San Jose, *H. E. Ahles 4074*) counties.
This rare species of the Illinois sand prairies turns blackish when dry. *Sisyrinchium montanum* var. *crebrum*, which also turns black upon drying, has wider

85. *Sisyrinchium campestre* (Blue-eyed Grass). *a.* Habit, X¼. *b.* Capsule, X3¾. *c.* Seed, X10.

86. *Sisyrinchium mucronatum* (Blue-eyed Grass). X¼.

stems, longer capsules, and bracts united for 2–6 mm above the base.

6. Sisyrinchium montanum Greene, Pittonia 4:33. 1899. *Fig. 87.*

Plants pale green or glaucescent or bright green drying pale or blackish; leaves 1–3 mm broad; stems to 60 cm tall, broadly or narrowly winged, 3–4 mm wide; spathe sessile, 1, the margin united for 2–6 mm above base; perianth violet; capsule subgloboid, whitish, greenish, pale brown, or stramineous, sometimes becoming black at maturity, 4–6 mm long.

Two variations occur in Illinois, both of which are extremely rare. These two taxa have sometimes been treated in the past as separate species, but the differences between the two do not appear to be specifically significant.

These two variants may be distinguished by the following key:

a. Plants pale green or glaucous, drying pale; capsule whitish, greenish, or stramineous, even at maturity ______ 6a. *S. montanum* var. *montanum*

a. Plants bright green, drying blackish; capsule greenish or pale brown at first, becoming dark at maturity __ 6b. *S. montanum* var. *crebrum*

6a. Sisyrinchium montanum Greene var. **montanum**

COMMON NAME: Blue-eyed Grass.

HABITAT: Sandy prairies.

RANGE: Quebec to British Columbia, south to New Mexico, Illinois, and North Carolina.

ILLINOIS DISTRIBUTION: Rare, known only from Cook and Lake counties; not collected since 1908.

The generally pallid aspect of the leaves, stems, and capsules makes this taxon rather strikingly different in appearance from the following variety.

87. *Sisyrinchium montanum* (Blue-eyed Grass). X¼.

6b. Sisyrinchium montanum Greene var. **crebrum** Fern. Rhodora 48:159. 1946.

COMMON NAME: Blue-eyed Grass.

HABITAT: Fields.

RANGE: Newfoundland to Ontario, south to Indiana and Virginia.

ILLINOIS DISTRIBUTION: Very rare; a single collection from Kankakee County.

This taxon has sometimes in the past been erroneously called S. *angustifolium*, but Ward (1968) has presented evidence to show why the binomial S. *angustifolium* does not apply to this taxon.

BURMANNIACEÆ—BURMANNIA FAMILY

Green or non-green, mostly tropical herbs; perianth parts 6, united; stamens 3 or 6; carpels 3, united, inferior.

Only the following genus occurs in Illinois.

1. *Thismia* GRIFF.

Non-green saprophyte; leaves minute, cauline; flowers regular, perfect; perianth parts 6; stamens 6; ovary 1, inferior.

Only the following species occurred in Illinois.

1. Thismia americana N. E. Pfeiffer, Bot. Gaz. 57:123. 1914. *Fig. 88.*

Sarcosiphon americanus (N. E. Pfeiffer) Schlecht. Notizbl. Bot. Gart. Mus. Berlin 71:39. 1921.

Stem subterranean, bearing a single partly subterranean flower; leaves minute, scaly; flower white or greenish, 8–15 mm long, with 6 spreading lobes, the throat closed by an annulus; stamens pendent from the annulus.

HABITAT: Wet prairies.

RANGE: Known only from Chicago, Illinois; now apparently extinct.

The discovery of this species, whose nearest relative is in New Zealand and Tasmania, was in August, 1912, by a class from the University of Chicago under Norma Pfeiffer. It has been repeatedly sought again elsewhere, but with no success. The original area is covered by industrial sites today. The nearest relationship is with *Thismia rodwayi* of New Zealand and Tasmania.

88. *Thismia americana* (Thismia). X3.

Order Orchidales

Only the following family comprises the order.

ORCHIDACEÆ–ORCHID FAMILY

Flowers perfect, irregular, greatly modified; perianth parts (5–) 6, in 2 series; stamens 1 or 2, united with the style to form a central column; ovary 1, inferior, 1-celled; seeds minute.

This family is considered by some to be the largest in the world with over 20,000 species thus far known. Most of these are tropical. Some species entirely lack chlorophyll. Most all species are mycorrhizal. The stamens are united to the style to form a central column. Pollen frequently coheres in waxy masses known as pollinia.

The Illinois genera of orchids traditionally fall into four tribes on the basis of staminal structures.

Tribe Cypripedieae, represented by *Cypripedium,* is distinguished by the presence of two fertile anthers.

Tribe Ophrydeae, with a single persistent fertile anther, is composed of the genera *Orchis* and *Habenaria.*

The majority of the Illinois genera possess a single fertile anther which is shed very early. Some of these have granular or powdery pollen and are assigned to Tribe Neottieae; others have more or less waxy pollen and belong to Tribe Epidendreae. Illinois genera in Tribe Neottieae are *Pogonia, Isotria, Triphora, Calopogon, Epipactis, Spiranthes,* and *Goodyera;* those in Tribe Epidendreae are *Corallorhiza, Malaxis, Liparis, Aplectrum, Tipularia,* and *Hexalectris.*

Several references are invaluable to the lover and student of orchids. The more important ones for Illinois are by Correll (1950), Ames (1924), Case (1964), and Winterringer (1967).

KEY TO THE GENERA OF **Orchidaceae** IN ILLINOIS

1. Plants with green leaves at flowering time.
   2. Leaves whorled ---------------------------------- 10. *Isotria*
   2. Leaves alternate or basal.
      3. Lip inflated, sac-like, at least 18 mm long; anthers 2 -- 1. *Cypripedium*
      3. Lip flat or, if inflated, less than 18 mm long; anther 1.
         4. Leaves 1 or 2, basal (leaf 1–2 and cauline in *Malaxis*).

5. Flowers totally pink, resupinate; leaf generally 1, linear to linear-lanceolate__________________2. *Calopogon*
5. Flowers greenish-yellow, madder-purple, or white and pink, not resupinate; leaves generally 2, lanceolate to elliptic to ovate to orbicular.
    6. Flowers spurred, usually fragrant; anther persistent, difficult to detach.
        7. Flowers white and pink; sepals and petals united to form a hooded structure (galea) behind the column___________________________3. *Orchis*
        7. Flowers greenish-yellow; sepals and petals not forming a galea________________4. *Habenaria*
    6. Flowers spurless, not fragrant; anther easily and soon detachable_____________________5. *Liparis*

4. Leaves more than 2, basal or cauline or, if only 1–2, then the leaf cauline.
    8. Leaves 1–2, cauline____________________6. *Malaxis*
    8. Leaves 3 or more, basal or cauline or both.
        9. All leaves basal.
            10. Leaves green throughout; lip not sac-like____ ____________________________7. *Spiranthes*
            10. Leaves conspicuously marked with white; lip sac-like______________________8. *Goodyera*
        9. At least 1 or more cauline leaves present (basal leaves also may be present).
            11. Flowers spurred; anther persistent, not easily detached______________________4. *Habenaria*
            11. Flowers not spurred; anther easily detached.
                12. Leaves both basal and cauline; flowers 1–2, terminal, or numerous, sessile and spicate.
                    13. Cauline leaf 1; sepals and petals free; flowers 1–2, pink, at least 15 mm long ________________________9. *Pogonia*
                    13. Cauline leaves 2 or more; upper sepal united with petals; flowers numerous, white, at most 10 mm long__________ _____________________7. *Spiranthes*
                12. Leaves all cauline; flowers numerous, pedicellate, racemose.

14. Flowers white (rarely pink), nodding, 1.0–1.5 cm long, the lip not sac-like; sepals and petals lanceolate; leaves weak, oval_____________11. *Triphora*
14. Flowers green and purple, ascending, well over 1.5 cm long, the lip sac-like at the base; sepals and petals ovate; leaves firm, lanceolate to ovate______ ____________________12. *Epipactis*

1. Plants without green leaves, at least at flowering time.
   15. Green leaves never produced; stem brown or purple, with colored sheaths; rhizomes coralline.
      16. Lip without longitudinal ridges; flowers at most 1 cm long, brown, yellow-green, or white spotted with purple__ _________________________________13. *Corallorhiza*
      16. Lip with 5–6 longitudinal ridges; flowers 1.7–2.3 cm long, madder-purple________________________14. *Hexalectris*
   15. Green leaves produced, but these usually absent at flowering time; plants with fleshy roots or with corms or tubers connected in a series.
      17. Leaves more than 1; plants with fleshy roots, but without connected tubers; flowers whitish__________7. *Spiranthes*
      17. Leaf 1; plants with connected tubers; flowers yellow or brown or madder-purple.
         18. Leaves green on both sides; flowers not spurred_____ _______________________________15. *Aplectrum*
         18. Leaves purple beneath; flowers spurred__16. *Tipularia*

1. *Cypripedium* L. – Lady's-slipper Orchid

Perennials with thick, fibrous roots; stems pubescent; leaves basal or cauline, large; flowers 1–3, large; sepals 3, spreading, the lower 2 usually united; petals spreading; lip sac-like, inflated; anthers 2; staminodium 1, petaloid; fruit a capsule.

KEY TO THE SPECIES OF Cypripedium IN ILLINOIS

1. Lip cleft down the middle, pink; leaves 2, basal______1. *C. acaule*
1. Lip not cleft, yellow, white, or with pink or purple streaks; leaves more than 2, cauline.
   2. Sepals and petals acuminate, yellow or greenish-yellow or streaked with purple.
      3. Lip yellow______________________________2. *C. calceolus*
      3. Lip white, occasionally marked with purple streaks.

4. Lip white; staminodium narrowly triangular ---------- 3. *C.* × *andrewsii*

4. Lip white, streaked with purple; staminodium oblong ---------- 4. *C. candidum*

2. Sepals and petals obtuse, white ---------- 5. *C. reginae*

1. Cypripedium acaule Ait. Hort. Kew. 3:303. 1789. *Fig.* 89.

Leaves 2, basal, broadly elliptic, glandular-pubescent, to 25 cm long, to 15 cm broad; scape to 50 cm tall, without bracts; flower 1, terminal; upper sepal narrowly lanceolate, greenish-brown to greenish-purple, to 5 cm long, to 1.5 cm broad; lateral sepals united, proportionately broader, to 4 cm long, to 2.5 cm broad, greenish-brown to greenish-purple; petals lanceolate, greenish-brown, pubescent within, scarcely twisted, to 5 cm long, to 1.5 cm broad; lip pink or rose, 3–7 cm long, drooping, with a deep cleft down the middle, pubescent on the inner surface; staminodium round; capsule to 5 cm long.

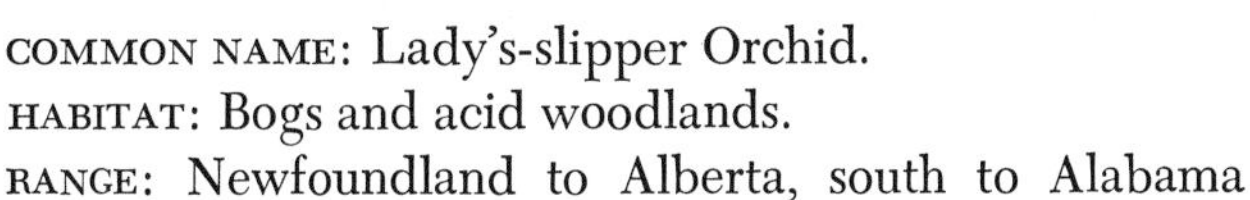

COMMON NAME: Lady's-slipper Orchid.

HABITAT: Bogs and acid woodlands.

RANGE: Newfoundland to Alberta, south to Alabama and Georgia.

ILLINOIS DISTRIBUTION: Very rare; known only from Cook County (Glencoe, May 20, 1908, *W. B. Day s.n.*; Elk Grove, May 3, 1942, *G. Pearsall* 7329).

This most beautiful orchid flowers from late May to early July. It is our only species of lady's-slipper with the leaves confined to the base of the plant and with the lip deeply cleft. Pepoon (1927) reports a collection by Brennan from Hog Island in Cook County, but this specimen has not been found during research for this book.

2. Cypripedium calceolus L. Sp. Pl. 951. 1753.

Leaves several, cauline, ovate-lanceolate to ovate, acute, to 20 cm long, to 12 cm broad; stem to 70 cm tall; flowers 1–2, terminal, each subtended by a foliaceous bract, to 10 cm long, to 5 cm broad; sepals greenish-yellow to purplish, the upper broadly lanceolate, to 7 cm long, 2–3 cm broad, the lower two usually connate; lateral petals purplish, linear-lanceolate, twisted, to 10 cm long, up to 1 cm broad; lip yellow, usually faintly purple-veined or -spotted, 2–6 cm long; staminodium ovoid; capsule to 5 cm long.

89. *Cypripedium acaule* (Lady's-slipper Orchid). *a.* Habit, X¼. *b.* Capsule, X¼. *c.* Habit (shaded), X1/24.

Two generally recognizable varieties may be found in Illinois. A discussion of these may be found in Fuller (1933).

a. Plants of bogs and marshes; flowers fragrant; lateral petals 3.5–5.0 cm long; staminodium truncate or tapering at base; sepals usually purplish______________________2a. *C. calceolus* var. *parviflorum*
a. Plants of woodlands; flowers not fragrant; lateral petals 5–8 cm long; staminodium rounded or subcordate at base; sepals usually yellow-green___________________2b. *C. calceolus* var. *pubescens*

2a. Cypripedium calceolus L. var. **parviflorum** (Salisb.) Fern. Rhodora 48:4. 1946. *Fig.* 90.

Cypripedium parviflorum Salisb. Trans. Linn. Soc. 1:77. 1791.
Plants to 50 cm tall; stem with 3–5 leaves; largest leaf at most 9 cm broad; flowers fragrant; sepals purple, upper 3–5 cm long; lateral petals 3.5–5.0 cm long; lip 2–4 cm long; staminodium truncate or tapering at base; 2n = 20 (Carlson, 1945, as *C. parviflorum*).

COMMON NAME: Small Yellow Lady's-slipper Orchid.
HABITAT: Bogs and marshes.
RANGE: Newfoundland to British Columbia, south to Washington, Texas, and Georgia.
ILLINOIS DISTRIBUTION: Rare; restricted to the northern counties.

The characters given in the key are the most reliable in separating this variety from var. *pubescens*. Additional characters in the above description sometimes overlap those in var. *pubescens*. The fragrant flowers appear from late May to late June.

2b. Cypripedium calceolus L. var. **pubescens** (Willd.) Correll, Bot. Mus. Leafl. Harv. Univ. 7:14. 1938. *Fig.* 91.

Cypripedium pubescens Willd. Hort. Berol. 4:143. 1805.
Cypripedium parviflorum var. *pubescens* (Willd.) Knight, Rhodora 8:93. 1906.
Plants to 70 cm tall; stem with 4–6 leaves; largest leaf to 12 cm broad; flowers not fragrant; sepals yellow-green, the upper 4–7 cm long; lateral petals 5–8 cm long; lip 3–5 cm long; staminodium rounded or subcordate at base.

90. *Cypripedium calceolus* var. *parviflorum* (Small Yellow Lady's-slipper Orchid). *a*. Habit, X¼. *b*. Capsule, X¼. *c*. Habit (shaded), X1/24.

91. *Cypripedium calceolus* var. *pubescens* (Yellow Lady's-slipper Orchid). *a*. Habit, X¼. *b*. Capsule, X¼. *c*. Habit (shaded), X1/24.

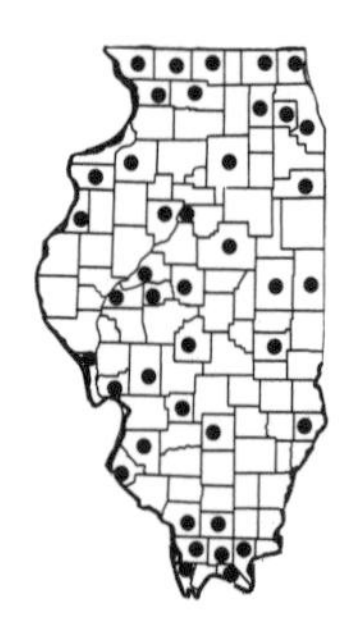

COMMON NAME: Yellow Lady's-slipper Orchid.
HABITAT: Moist or dry woodlands.
RANGE: Nova Scotia to Minnesota, south to Alabama and Georgia.
ILLINOIS DISTRIBUTION: Not common; scattered throughout the state.

Although not common, this variety is more plentiful than var. *parviflorum*. The flowers are produced from mid-April to mid-May. Willdenow originally considered this taxon distinct from both the Eurasian *C. calceolus* and the North American *C. parviflorum*. Enough intergradation occurs between the characters of these taxa that most orchid experts today treat them as varieties of one species. The typical Eurasian variety is generally even smaller than our var. *parviflorum*.

3. **Cypripedium × andrewsii** A. M. Fuller, Rhodora 34:100. 1932. *Fig.* 92.

Cypripedium candidum × *parviflorum* A. M. Fuller, Bull. Publ. Mus. Milwaukee 14:72. 1933.

Leaves several, cauline, lance-ovate, acute, to 7 cm broad; stem to 50 cm tall; flowers 1–2, terminal, slightly fragrant, each subtended by a foliaceous bract; sepals purple, the upper ovate-lanceolate, 2.5–3.7 cm long, the lower two connate; lateral petals purple or greenish-purple, lanceolate, twisted, 3–4 cm long; lip creamy-white, 2.0–2.5 cm long; staminodium narrowly triangular.

COMMON NAME: Lady's-slipper Orchid.
HABITAT: Low, springy situations.
RANGE: Wisconsin and Illinois.
ILLINOIS DISTRIBUTION: Very rare; known only from Woodford County (near Spring Bay, *V. H. Chase 4024*).

This beautiful plant, which flowers in June, is a hybrid between *C. calceolus* var. *parviflorum* and *C. candidum*. In most of its characters, it is intermediate between these two taxa. Its resemblance, however, is more with *C. candidum*.

92. *Cypripedium* × *andrewsii* (Lady's-slipper Orchid). *a.* Habit, X¼. *b.* Habit (shaded), X1/24.

4. **Cypripedium candidum** Muhl. ex Willd. Sp. Pl. 4:142. 1805. *Fig. 93.*

Leaves several, cauline, stiff, erect, lanceolate to elliptic, acute, to 15 cm long, to 4 cm broad, the lowest reduced to sheathing scales; stem to 40 cm tall, glandular-pubescent; flower 1, terminal, slightly fragrant, subtended by an erect foliaceous bract to 6 cm long, to 2 cm broad; sepals greenish-yellow striped with purple, the upper narrowly lanceolate, 2.0–3.5 cm long, to 1 cm broad, the lower two connate; lateral petals greenish-yellow striped with purple, linear-lanceolate, long-tapering at the apex, twisted, to 5 cm long, to 0.4 cm broad; lip white, striped with purple, waxy-looking, to 2.5 cm long; staminodium oblongoid; capsule erect, to 3 cm long.

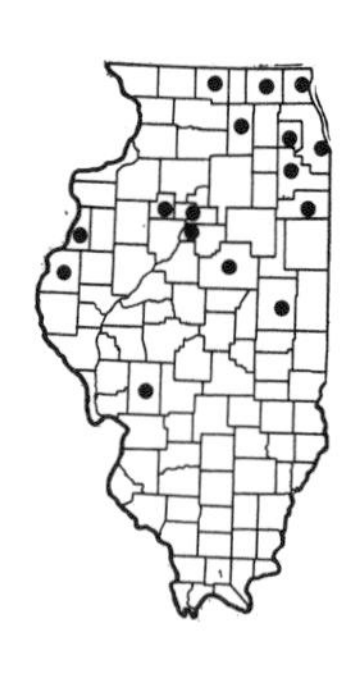

COMMON NAME: White Lady's-slipper Orchid.

HABITAT: Bogs, swamps, and wet prairies.

RANGE: New York to South Dakota, south to Nebraska, Missouri, and Pennsylvania.

ILLINOIS DISTRIBUTION: Very rare; specimens have been collected from 16 counties in the northern half of the state.

This species, along with the Yellow Lady's-slipper and the Showy Lady's-slipper, were not infrequent in the state before 1900. Indiscriminate collecting, however, has made all these species rare in Illinois to the point of extinction.

The White Lady's-slipper flowers from mid-May to mid-June.

5. **Cypripedium reginae** Walt. Fl. Carol. 222. 1788. *Fig. 94.*

Cypripedium spectabile Salisb. Trans. Linn. Soc. 1:78. 1791.

Leaves several, cauline, pubescent on the margins and veins, broadly elliptic to ovate, short-acuminate, to 30 cm long, to 15 cm broad; stem to 90 cm tall, hirsute; flowers 1–3, terminal, subtended by a foliaceous bract to 10.5 cm long; sepals white, waxy-looking, obtuse, the upper suborbicular, to 5 cm long, to 3.5 cm broad, at length arching over the lip, the lower two connate; lateral petals white, obtuse, oblong, to 4 cm long, flat; lip white but strongly marked with pink or rose, 3.0–4.5 cm long; capsule to 4.5 cm long.

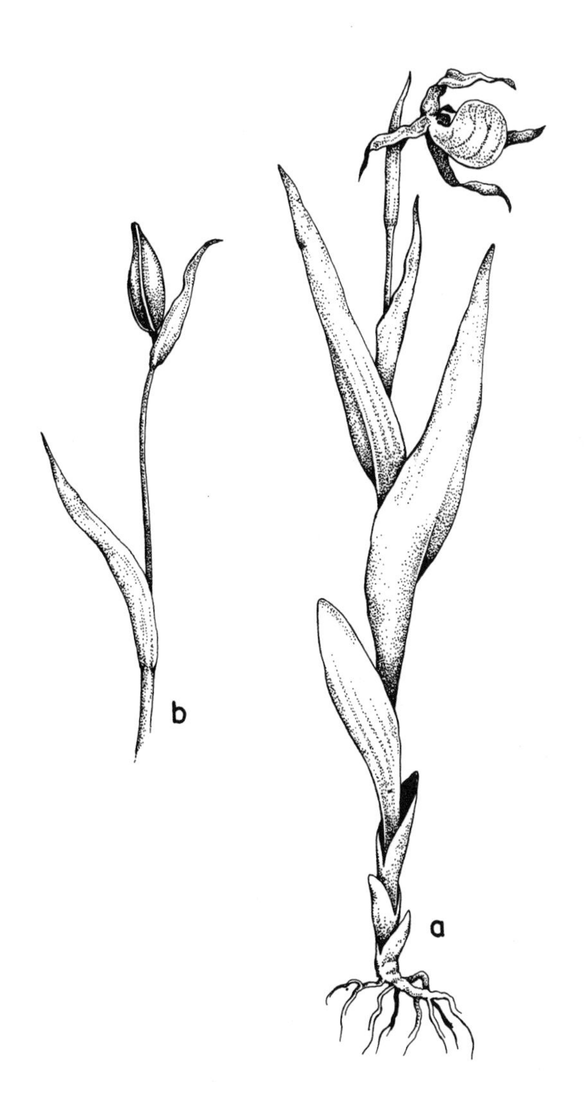

93. *Cypripedium candidum* (White Lady's-slipper Orchid). *a.* Habit, X¼. *b.* Capsule, X¼.

94. *Cypripedium reginae* (Showy Lady's-slipper Orchid). *a.* Habit, X¼. *b.* Capsule, X¼. Habit (shaded), X1/24.

COMMON NAME: Showy Lady's-slipper Orchid.
HABITAT: Bogs and other low, springy places.
RANGE: Newfoundland to Manitoba, south to North Dakota, Missouri, and Georgia.
ILLINOIS DISTRIBUTION: Rare; known from 12 counties in the northern one-third of the state.

This is undoubtedly the most beautiful orchid in Illinois. Its beauty, however, has led to its near extinction. The stems and leaves are so harshly pubescent that persons handling the plant sometimes break out in a rash.

The flowers are borne in June and early July.

Until Pepoon's publication in 1909 on the cliff flora of JoDaviess County, this species had been known to Illinois botanists as *C. spectabile.*

2. *Calopogon* L. C. RICHARD – Grass Pink Orchid

Perennial from a solid bulb; stems slender, glabrous, with a single nearly basal leaf and 1 or 2 sheathing scale-like leaves; flowers racemose, resupinate; sepals 3, free, spreading; petals free, spreading, the lip clawed, bearded on the inner face and papillose at the apex; column winged near the apex; anther 1, with 4 pollen masses; fruit a capsule.

Only the following species occurs in Illinois.

1. Calopogon tuberosus (L.) BSP., Prel. Cat. N. Y. 1888. *Fig.* 95.

Limodorum tuberosum L. Sp. Pl. 950. 1753.
Limodorum pulchellum Salisb. Prod. Hort. Chap. Allerton 8. 1796.
Calopogon pulchellus (Salisb.) R. Br. ex Ait. Hort. Kew. 5:204. 1813.

Leaf usually 1, linear to linear-lanceolate, to 50 cm long, to 5 cm broad, long-sheathing at the base; stem to 75 cm tall; flowers 2–15, resupinate, 2.5–4.5 cm broad, subtended by a bract to 1 cm long; pedicels 3–9 mm long; sepals broadly elliptic, acute, concave, to 2.5 cm long, to 1 cm broad, pink or pinkish-purple; petals broadly lanceolate, subacute or obtuse, to 2.5 cm long, to 1 cm broad, pink or pinkish-purple; lip clawed below, expanded above, three-lobed, white-pubescent on the inner face, pink or pinkish-purple tipped with yellow, to 2 cm long, to nearly 1 cm broad; capsule angular, to 2.5 cm long, strongly ribbed.

95. *Calopogon tuberosus* (Grass Pink Orchid). *a.* Habit, X¼. *b.* Flower, X1. *c.* Habit (in fruit), X⅕.

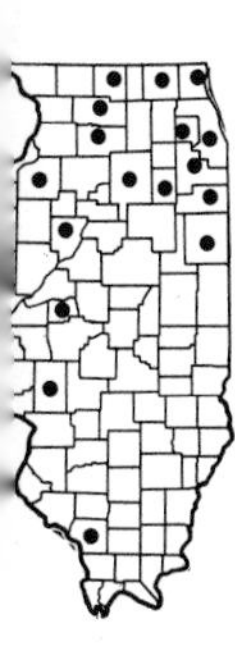

COMMON NAME: Grass Pink Orchid.
HABITAT: Wet prairies.
RANGE: Newfoundland to Minnesota, south to Texas and Florida.
ILLINOIS DISTRIBUTION: Not common; recorded from 18 counties in the northern half of the state, plus Jackson County.
The large pink resupinate flower and the single grass-like leaf readily distinguish this species which flowers from mid-May until early July.

3. *Orchis* L. – Orchis

Acaulescent perennial from short rhizomes with fleshy roots; leaves 2, basal; scape bearing a terminal, bracteate raceme; flowers fragrant; sepals and petals connivent to form a hood behind the column; lip large, spurred; anther 1, persistent, difficult to detach; fruit a capsule.

Only the following species occurs in Illinois.

1. Orchis spectabilis L. Sp. Pl. 939. 1753. *Fig. 96.*

Leaves 2, basal, broadly elliptic, obtuse to subacute, rather thick, shining, glabrous, to 20 cm long, to 10 cm broad; scape to 15 cm tall, 4- to 5-angled, glabrous; inflorescence racemose, 3- to 10-flowered; flowers fragrant, subtended by foliaceous bracts to 6 cm long, to 1.5 cm broad; sepals and petals pink, ovate-oblong, to 18 mm long, to 6 mm broad, connivent to form a hood; lip broadly ovate, white, to 15 mm long, to 12 mm broad, stout spur up to 2 cm long; capsule to 2.5 cm long, ribbed; 2n = 42 (Humphrey, 1932).

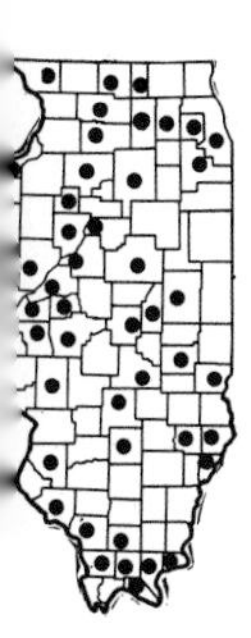

COMMON NAME: Showy Orchis.
HABITAT: Rich woodlands.
RANGE: Quebec to Ontario, south to Kansas, Alabama, and Georgia.
ILLINOIS DISTRIBUTION: Not common; scattered throughout the state.
This is one of the loveliest and most daintily scented wild flowers in Illinois. The variegated flowers are produced from late April to early June.

96. *Orchis spectabilis* (Showy Orchid). *a.* Habit, X¼. *b.* Flower, X½. *c.* Lip petal, X1. *d.* Fruiting cluster, X½. *e.* Habit (shaded), X1/24.

4. *Habenaria* WILLD. – Fringed Orchis

Caulescent (rarely acaulescent) perennials from tubers, rhizomes, or fleshy roots; leaves 2-several, alternate or basal; inflorescence spicate or racemose, bracteate; flowers usually fragrant; sepals and petals separate, similar; lip spurred; anther 1, persistent, difficult to detach; fruit a capsule.

This is the largest genus of Orchidaceae in Illinois. Many of the species are exceedingly rare and may now be extinct in the state.

KEY TO THE TAXA OF Habenaria IN ILLINOIS

1. Leaves 2, basal, suborbicular; lip elongated, entire, 8–20 mm long, the spur 13–40 mm long.
    2. Lip linear, 12–20 mm long; flowers greenish-white, with pedicels 5–13 mm long; stem bracted________________1. *H. orbiculata*
    2. Lip triangular-elongated, 8–12 mm long; flowers yellow-green, sessile; stem without bracts____________________2. *H. hookeri*
1. Leaves 1-several, cauline, linear to lanceolate to oval; lip lobed, toothed, erose, or fringed, if entire and elongated, then the lip 4–8 mm long and the spur 2–12 mm long.
    3. Lip entire, erose, 2- to 3-toothed, or very shallowly 3-lobed, with the lobes entire; spur 2–12 mm long; flowers sessile.
        4. Leaf 1; inflorescence 2–6 cm long; all bracts shorter than the flowers; spur 8–12 mm long; roots slender___3. *H. clavellata*
        4. Leaves 2-several; inflorescence 5–30 cm long; at least the lower bracts longer than the flowers; spur 2–8 mm long; rootstocks tuberous.
            5. Sepals at least twice as long as the spur, the spur 2–3 mm long; flowers greenish, tinged with purple; lip 2- to 3-toothed___________________4. *H. viridis* var. *bracteata*
            5. Sepals shorter than to as long as the spur, the spur 3–8 mm long; flowers greenish-white, greenish-yellow, or white; lip entire or erose.
                6. Lip erose, with a tubercle borne near the summit_____ ____________________________________5. *H. flava*
                6. Lip entire, without a tubercle.
                    7. Flowers greenish-white, faintly odorous; lip gradually broadened at the base_______6. *H. hyperborea*
                    7. Flowers creamy-white, with a spicy fragrance; lip abruptly broadened at the base______7. *H. dilatata*
    3. Lip fringed or deeply 3-lobed and toothed; spur 12 mm long or longer; flowers pedicellate.

8. Lip simple, fringed.
   9. Flowers orange; leaves lanceolate, some or all over 2 cm broad________________________________8. *H. ciliaris*
   9. Flowers white; leaves linear to linear-lanceolate, less than 2 cm broad______________________9. *H. blephariglottis*
8. Lip deeply 3-lobed, the lobes fringed, long-toothed, or erose.
   10. Flowers yellow-green or white; lobes of the lip fringed.
      11. Flowers yellow-green or greenish-white; sepals 4.5–7.0 mm long; spur 14–20 mm long____10. *H. lacera*
      11. Flowers white; sepals 7–12 mm long; spur 20–48 mm long____________________11. *H. leucophaea*
   10. Flowers reddish-purple; lobes of the lip long-toothed or erose.
      12. Lip shallowly erose, 13–22 mm long, the terminal lobe notched_________________12. *H. peramoena*
      12. Lip long-toothed, 9–13 mm long, the terminal lobe not notched____________________13. *H. psycodes*

1. Habenaria orbiculata (Pursh) Torr. Compend. Fl. N. & Mid. States 318. 1826. *Fig.* 97.

Orchis orbiculata Pursh, Fl. Am. Sept. 588. 1814.

Rootstocks fleshy, tuberous; leaves 2, basal, rather thin, more or less orbicular, 8–14 cm long, nearly as broad; stem to 60 cm tall; inflorescence rather sparse, 5–20 cm long; flowers greenish-white; pedicels 5–13 mm long; sepals somewhat longer than the petals, 8–12 mm long; lip linear, 12–20 mm long, entire, deflexed; spur 18–40 mm long.

COMMON NAME: Round-leaved Orchid.

HABITAT: Moist woodlands.

RANGE: Labrador to Alaska, south to Washington, Illinois, and Georgia.

ILLINOIS DISTRIBUTION: Very rare; only one collection seen (Kane Co.: Elgin, *G. Vasey s.n.*).

This and the following species are the only species of *Habenaria* in Illinois with a pair of basal leaves only. *Habenaria orbiculata* is distinguished from *H. hookeri* by its pedicellate flowers, its narrower, longer lip petal, and its thinner leaves.

The flowers are borne from mid-July to mid-August.

97. *Habenaria orbiculata* (Round-leaved Orchid). *a.* Habit, X⅙. *b.* Flower, X¾. *c.* Cluster of fruits, X⅓. *d.* Habit (shaded), X1⁄24.

2. **Habenaria hookeri** Torr. ex Gray, Ann. Lyc. N. Y. 3:228. 1835. *Fig.* 98.

Habenaria hookeriana Torr. ex Gray, Ann. Lyc. N. Y. 3:229. 1835.

Rootstocks tuberous; leaves 2, basal, lying flat on the ground, thick, more or less orbicular, rarely broadly elliptic, 7–12 cm long, nearly as broad, glabrous; stem to 40 cm tall, glabrous; inflorescence rather sparse, 5–25 cm long; flowers yellow-green, sessile, each subtended by a bract to 1.5 cm long; sepals somewhat longer than the petals, 6–11 mm long, to 6 mm broad; lip triangular, 8–12 mm long, to 5 mm broad, entire, deflexed; spur 13–25 mm long.

COMMON NAME: Hooker's Orchid.

HABITAT: Moist woodlands.

RANGE: Nova Scotia to Minnesota, south to Iowa and West Virginia.

ILLINOIS DISTRIBUTION: Rare; known only from Lake and Cook counties in extreme northeastern Illinois.

Flowering time for this species is early June to late July.

3. **Habenaria clavellata** (Michx.) Spreng. Syst. 3:689. 1826. *Fig.* 99.

Orchis clavellata Michx. Fl. Bor. Am. 2:155. 1803.

Orchis tridentata Muhl. ex Willd. Sp. Pl. 4:41. 1805.

Habenaria tridentata (Muhl.) Hook. Exot. Fl. 2:pl. 81. 1825.

Gymnadeniopsis clavellata (Michx.) Rydb. in Britt. Man. 293. 1901.

Rootstocks rather slender, not tuberous; leaf 1, near the base of the plant, oblanceolate, glabrous, to 20 cm long, 1.5–3.0 cm broad; stem to 45 cm tall, glabrous; inflorescence slender, rather sparse, 2–6 cm long; flowers greenish-white or greenish-yellow, often twisted, each subtended by a bract, to 12 mm long; sepals and petals about equal, oval, 2–4 mm long, nearly as broad; lip broadly oblong, shallowly 3-lobed, 3–7 mm long; spur curved, slender, 8–12 mm long.

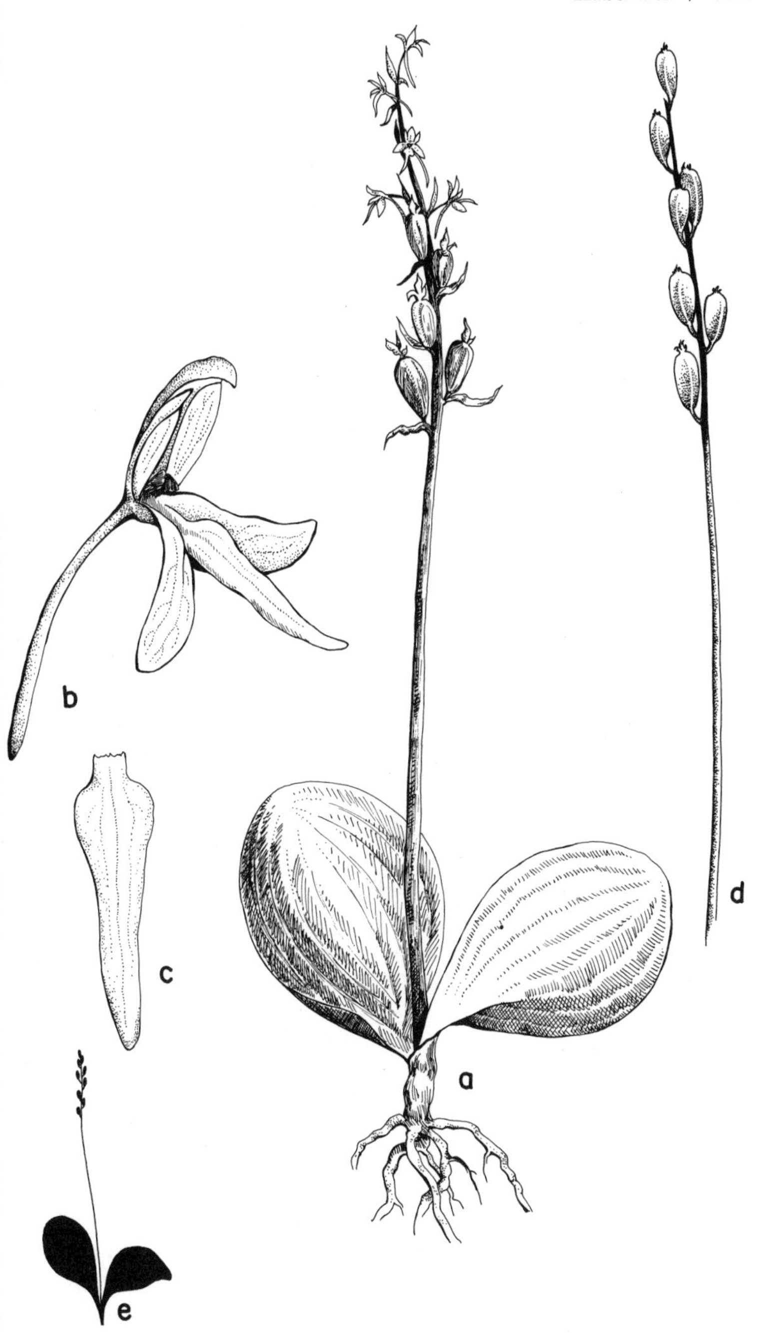

98. *Habenaria hookeri* (Hooker's Orchid). *a.* Habit, X¼. *b.* Flower, X1½. *c.* Lip, X2. *d.* Fruiting branch, X½. *e.* Habit (shaded), X1/24.

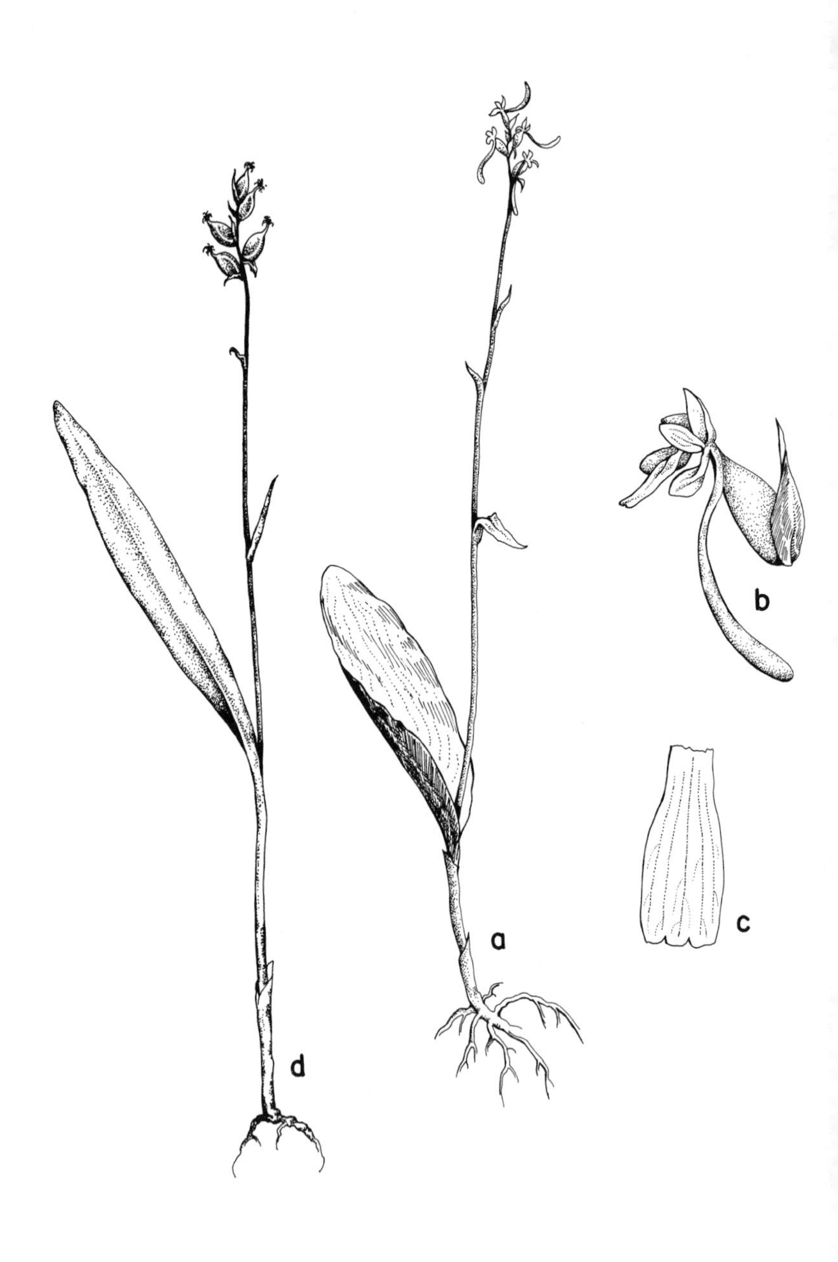

99. *Habenaria clavellata* (Wood Orchid). *a.* Habit, in flower, X¼. *b.* Flower, X1¾. *c.* Lip, X2½. *d.* Habit, in fruit, X¼.

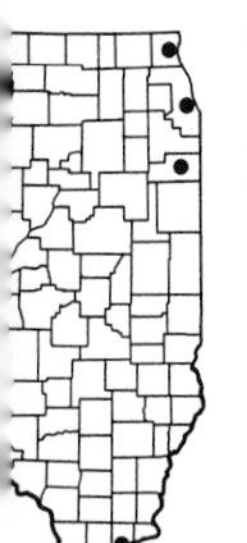

COMMON NAME: Wood Orchid; Green Orchid.
HABITAT: Moist, shaded areas.
RANGE: Newfoundland to Minnesota, south to Texas and Florida.
ILLINOIS DISTRIBUTION: Rare; known only from three counties in extreme northeastern Illinois and Cass County; also Pope County (low swampy woods, Massac Tower, *J. Schwegman;* Long Spring, *J. Schwegman*).

This is the smallest and most slender of the Habenarias in Illinois. Only one leaf is developed on each stem, but several reduced scale-like leaves may be found on the stem.

The flowers are produced in late July and early August.

4. **Habenaria viridis** (L.) R. Br. var. **bracteata** (Muhl.) Gray, Man. Bot. 5:500. 1867. *Fig. 100.*

Orchis bracteata Muhl. ex Willd. Sp. Pl. 4:34. 1805.
Habenaria bracteata (Muhl.) R. Br. in Ait. Hort. Kew. 5:192. 1813.
Platanthera bracteata (Muhl.) Torr. Fl. N. Y. 2:279. 1843.
Coeloglossum bracteatum Parl. Fl. Ital. 3:409. 1858.

Rootstocks tuberous; leaves several, cauline, oblanceolate to obovate, glabrous, to 15 cm long, to 5 cm broad; stem to 50 cm long, glabrous; inflorescence 5–20 cm long; flowers greenish marked with purple, sessile; bracts linear-lanceolate, conspicuous, all surpassing the flowers, to 5 cm long; sepals somewhat longer than the petals, 6–7 mm long; lip oblong, 2- to 3-toothed, 6–10 mm long, to 5 mm broad; spur 2–3 mm long, saccate.

COMMON NAME: Bracted Green Orchid.
HABITAT: Rich woodlands.
RANGE: Newfoundland to Alberta, south to Nebraska, Iowa, and South Carolina; Alaska.
ILLINOIS DISTRIBUTION: Rare; known from 13 counties in the northern one-third of the state.

This species is conspicuous because of the elongated bracts which subtend the flowers. The tiny spur is the shortest in the genus in Illinois. Flowers are produced from the last of May to the last of June.

Most workers recognize three varieties of *H. viridis*—var. *viridis,* from Eurasia, with short, broad bracts scarcely exceed-

100. *Habenaria viridis* var. *bracteata* (Bracted Green Orchid). *a.* Habit, X⅛. *b.* Flower, X2½. *c.* Capsule, X1¼.

ing the flowers and with a broad lip, var. *interjecta,* from northern North America and northeast Asia, with some bracts exceeding and some not surpassing the flowers, and var. *bracteata.* The characters used to separate these taxa are not definitive enough to warrant species status for the taxa.

5. **Habenaria flava** (L.) R. Br. ex Spreng. Syst. Veg. 3:691. 1826.

Orchis flava L. Sp. Pl. 942. 1753.

Perularia flava (L.) Farw. Ann. Rep. Parks Det. 11:54. 1900.

Rootstocks tuberous; leaves 1–5, cauline, linear-lanceolate to lanceolate, to 20 cm long, to 4.5 cm broad; stem to 60 cm tall; inflorescence 5–20 cm long, crowded or lax; flowers green or greenish-yellow, fragrant, subtended by longer or shorter bracts, sessile; sepals and petals about equal, 3–5 mm long, about half as wide; lip suborbicular to oblong, 4–6 mm long, erose, with a tubercle borne near the apex; spur slender, 3–10 mm long.

Two questionably distinguishable varieties occur in Illinois.

a. Bracts of lowest flowers equalling the flowers; inflorescence lax; lip suborbicular ________ 5a. *H. flava* var. *flava*

a. Bracts of lowest flowers much surpassing the flowers; inflorescence crowded; lip oblong ________ 5b. *H. flava* var. *herbiola*

5a. **Habenaria flava** (L.) R. Br. var. **flava** *Fig. 101.*

Platanthera flava (L.) Lindl. Gen. & Sp. Orch. 293. 1835.

Leaves 1–3; inflorescence lax; bracts of lowest flowers equalling the flowers; lip suborbicular.

COMMON NAME: Tubercled Orchid; Rein Orchid.

HABITAT: Low, shaded areas.

RANGE: Maryland to Missouri, south to Texas and Florida.

ILLINOIS DISTRIBUTION: Rare; known only from Massac, St. Clair, and Wabash counties.

This variety flowers from late June to mid-August.

101. *Habenaria flava* var. *flava* (Tubercled Orchid). *a*. Habit, X⅛. *b*. Flower, X2½. *c*. Lip, X3.

5b. Habenaria flava (L.) R. Br. var. **herbiola** (R. Br.) Ames & Correll, Bot. Mus. Leafl. Harv. Univ. 11:61. 1943. *Fig. 102.*

Orchis virescens Muhl. ex Willd. Sp. Pl. 4:37. 1805.
Habenaria herbiola R. Br. in Ait. Hort. Kew. 5:193. 1813.
Habenaria virescens (Muhl.) Spreng. Syst. 3:688. 1826.
Habenaria flava var. *virescens* (Muhl.) Fern. Rhodora 23:148. 1921.

Leaves 2–5; inflorescence crowded; bracts of the lowest flowers much surpassing the flowers; lip oblong.

COMMON NAME: Tubercled Orchid.
HABITAT: Moist woodlands.
RANGE: New Brunswick to Ontario, south to Missouri and North Carolina.
ILLINOIS DISTRIBUTION: Rare; restricted to nine counties in the northern one-third of the state. Except for shape of the lip, the characters separating these two taxa of *H. flava* are not strong.
The flowers appear in June and July.

6. Habenaria hyperborea (L.) R. Br. var. **huronensis** (Nutt). Farw. Papers Mich. Acad. Sci. 1:92. 1923. *Fig. 103.*

Orchis huronensis Nutt. Gen. 2:189. 1818.
Limnorchis huronensis (Nutt.) Rydb. in Britt. Man. Fl. N. U. S. 294. 1901.

Rootstocks tuberous; leaves several, cauline, lanceolate to oblanceolate, to 25 cm long, 2–5 cm broad; glabrous; stem to 80 cm tall, glabrous; inflorescence 5–30 cm long, usually crowded; flowers greenish to greenish-white, scarcely aromatic, sessile; sepals slightly longer than the petals, 3.5–10.0 mm long, to 4 mm broad; lip lanceolate, subacute at the apex, 4–10 mm long, to 3 mm broad, entire; spur 4–7 mm long.

102. *Habenaria flava* var. *herbiola* (Tubercled Orchid). *a.* Inflorescence, X⅛. *b.* Flowers, X3. *c.* Capsule, X⅝. *d.* Habit (shaded), X1/16.

103. *Habenaria hyperborea* var. *huronensis* (Green Orchid). *a*. Habit, X¼. *b*. Flower, X2½. *c*. Inflorescence, X⅓.

COMMON NAME: Green Orchid.
HABITAT: Swampy areas.
RANGE: Labrador to Alaska, south to Oregon, New Mexico, Illinois and Pennsylvania.
ILLINOIS DISTRIBUTION: Rare; known only from three northeastern counties and three north-central counties.
This taxon flowers from late June to early August.
The typical var. *hyperborea* is confined to extreme northern North America, Iceland, and northeastern Asia. It differs in its shorter inflorescences, smaller, aromatic flowers, and broader, blunter lip.

7. **Habenaria dilatata** (Pursh) Hook. Exot. Fl. t. 95. 1824. *Fig. 104.*

Orchis dilatata Pursh, Fl. Am. Sept. 2:588. 1814.
Limnorchis dilatata (Pursh) Rydb. in Britt. Man. Fl. N. U. S. 294. 1901.
Limnorchis fragrans Rydb. in Britt. Man. Fl. N. U. S. 294. 1901.
Habenaria fragrans (Rydb.) Niles, Bog-Trott. Orch. 253. 1904.

Rootstocks tuberous; leaves several, cauline, linear-lanceolate to lanceolate, to 20 cm long, 2–4 cm broad; stem to 80 cm tall; inflorescence 10–30 cm long, crowded; flowers white, strongly aromatic, sessile; sepals slightly longer than the petals, 4–5 mm long; lip suborbicular, abruptly widened at the base, 6–8 mm long, entire; spur 6–8 mm long.

COMMON NAME: White Orchis; Fragrant White Orchid.
HABITAT: Springy situations.
RANGE: Labrador to Alaska, south to California, New Mexico, Illinois, and Pennsylvania.
ILLINOIS DISTRIBUTION: Very rare; known only from McHenry County, and not collected since 1861.
The beautiful white, spicy-scented flowers make this a most attractive species. The flowers appear in late June or early July.

Habenaria × *media* Rydb., a hybrid between *H. dilatata* and *H. hyperborea* var. *huronensis,* has been attributed to Lake County by Pepoon (1927) and Fuller (1933), but no specimens

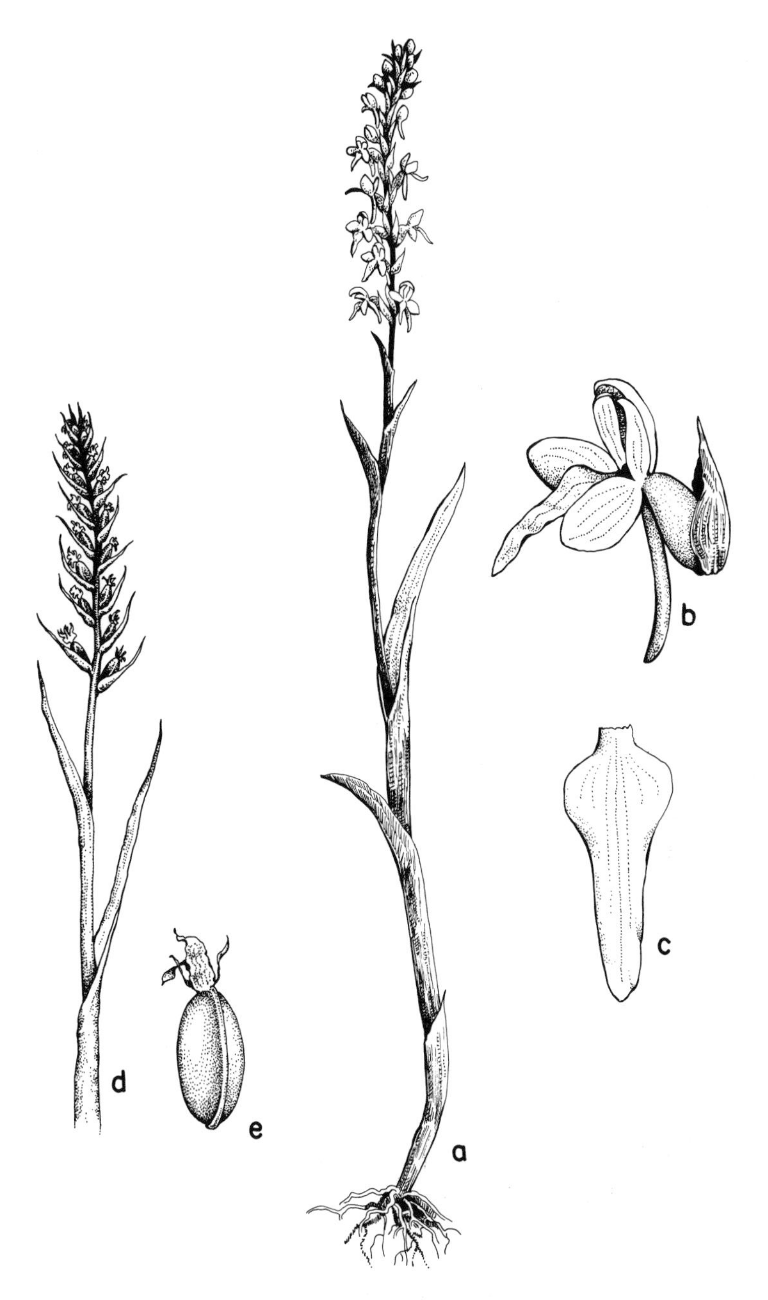

104. *Habenaria dilatata* (White Orchis). *a.* Habit, X⅛. *b.* Flower, X1¾. *c.* Lip, X2½. *d.* Fruiting branch, X⅛. *e.* Capsule, X½.

have been seen in the preparation of this book. In the hybrid, the flowers are greenish-yellow with the lip enlarged at the base.

8. **Habenaria ciliaris** (L.) R. Br. in Ait. Hort. Kew. 5:194. 1813. *Fig. 105.*

Orchis ciliaris L. Sp. Pl. 939. 1753.

Blephariglottis ciliaris (L.) Rydb. in Britt. Man. Fl. N. U. S. 296. 1901.

Rootstocks slender-tuberous; leaves 1–3, cauline, lanceolate, glabrous, to 25 cm long, 2–6 cm broad; stem to 90 cm tall, glabrous; inflorescence 5–15 cm long, rather lax; flowers orange, appearing pedicellate because of the pedicel-like ovary, each subtended by a bract to 2.5 cm long; sepals somewhat longer than the petals, 6–10 mm long, to 5 mm broad; lip broadly oblong, 10–15 mm long, to 3 mm broad, long-fringed; spur 15–30 mm long.

COMMON NAME: Yellow Fringed Orchid.

HABITAT: Low ground.

RANGE: Ontario to Illinois, south to Texas and Florida.

ILLINOIS DISTRIBUTION: Rare; apparently confined to Cook County.

This is the only orange-flowered *Habenaria* in Illinois; consequently it is the showiest, also.

The flowers are matured from mid-June to mid-August.

9. **Habenaria blephariglottis** (Willd.) Hook. Exot. Fl. 2:t. 87. 1824. *Fig. 106.*

Orchis blephariglottis Willd. Sp. Pl. 4:9. 1805.

Platanthera holopetala Lindl. Gen. & Sp. Orch. 291. 1835.

Rootstocks slender-tuberous; leaves 1–3, cauline, linear-lanceolate, to 20 cm long, 1–2 cm broad; stem to 75 cm tall; inflorescence 5–15 cm long, rather crowded; flowers white, appearing pedicellate; sepals somewhat longer than the petals, 5–9 mm long; lip broadly oblong, 8–11 mm long, fringed; spur 15–25 mm long.

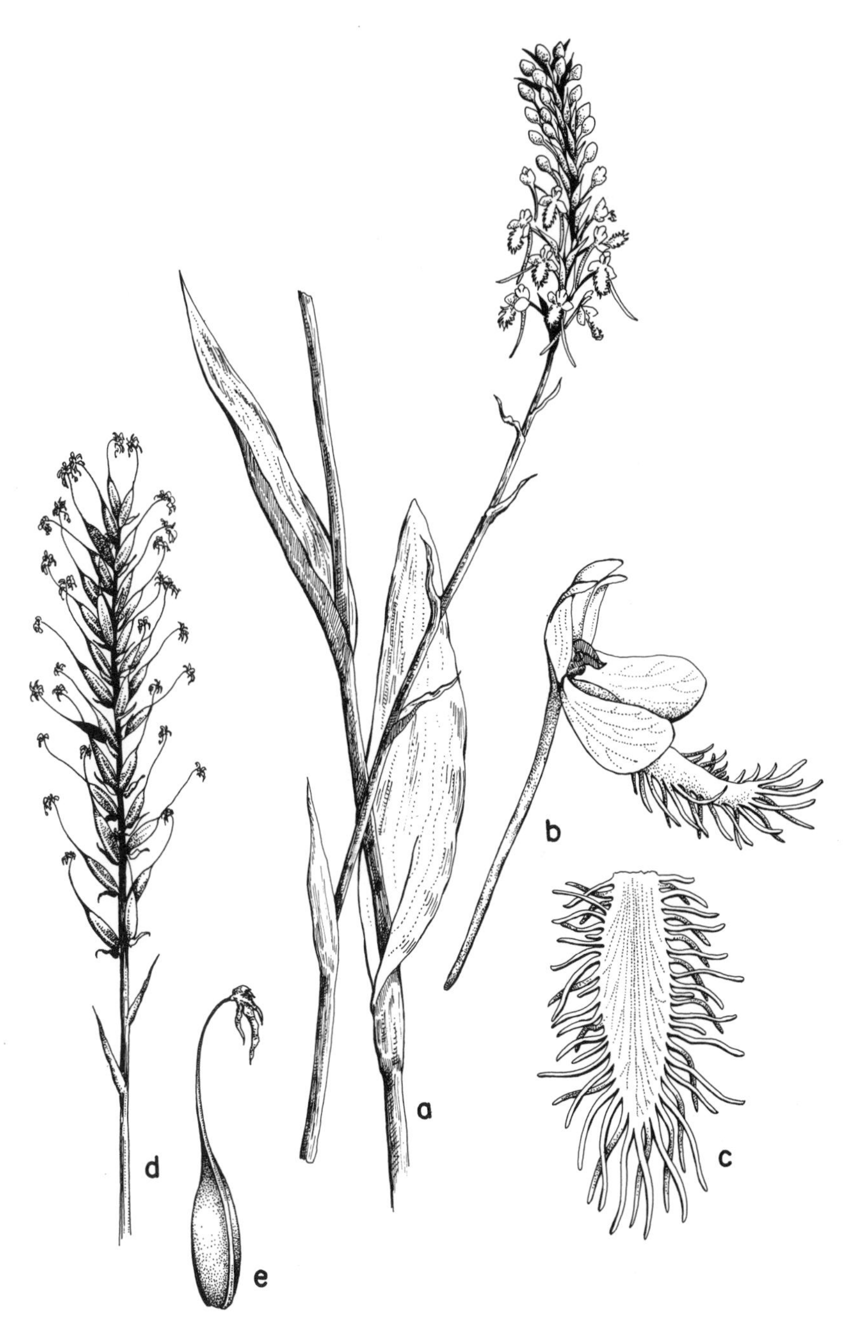

105. *Habenaria ciliaris* (Yellow Fringed Orchid). *a.* Inflorescence and leaves, X¼. *b.* Flower, X2½. *c.* Lip, X3½. *d.* Fruiting branch, X¼. *e.* Capsule, X½.

106. *Habenaria blephariglottis* (White Fringed Orchid). *a.* Inflorescence and leaves, X¼. *b.* Flower, X1¾. *c.* Lip, X1¾. *d.* Fruiting branch, X¼.

COMMON NAME: White Fringed Orchid.
HABITAT: Low areas.
RANGE: Newfoundland to Ontario, south to Illinois and South Carolina.
ILLINOIS DISTRIBUTION: Very rare; known only from Macon County (near Decatur).
The Macon County station is far to the south of the nearest known station for this handsome species. The flowers appear in late June or early July.

10. Habenaria lacera (Michx.) Lodd. Bot. Cab. 3:pl. 229. 1818. *Fig. 107.*

Orchis lacera Michx. Fl. Bor. Am. 2:156. 1803.
Platanthera lacera (Michx.) G. Don ex Sweet, Hort. Brit. 650. 1839.
Blephariglottis lacera (Michx.) Rydb. in Britt. Man. Fl. N. U. S. 296. 1901.

Rootstocks tuberous; leaves 4–9, cauline, lanceolate to oval, glabrous, to 18 cm long, to 4 cm broad; stem to 75 cm tall, glabrous; inflorescence 5–20 cm long, rather lax; flowers yellow-green, appearing short-pedicellate, subtended by bracts shorter than the flowers; sepals and petals about equal, 4.5–7.0 mm long, about half as broad; lip 3-lobed, with each lobe often 3-lobed, 10–16 mm long, each lobe long-fringed; spur 14–25 mm long, slender.

COMMON NAME: Green Fringed Orchid.
HABITAT: Swamps.
RANGE: Quebec to Ontario, south to Texas and Florida.
ILLINOIS DISTRIBUTION: Rare; restricted to the northern half of the state, plus Massac, Pope, and St. Clair counties.
The flowers open in June and July. Variation exists in size of leaves, sepals, petals, lip, and spur.

107. *Habenaria lacera* (Green Fringed Orchid). *a.* Inflorescence and leaves, X¼. *b.* Flower, X1½. *c.* Fruiting branch, X¼.

11. Habenaria leucophaea (Nutt.) Gray, Man. Bot. 502. 1867. *Fig. 108.*

Orchis leucophaea Nutt. Trans. Am. Phil. Soc. 5:161. 1834.
Platanthera leucophaea (Nutt.) Lindl. Gen. & Sp. Orch. 294. 1835.
Blephariglottis leucophaea (Nutt.) Rydb. in Britt. Man. Fl. N. U. S. 296. 1901.

Rootstocks tuberous; leaves 4–9, cauline, linear-lanceolate to lanceolate, glabrous, to 20 cm long, 2.0–3.5 cm broad; stem to 1 m tall, glabrous; inflorescence 8–20 cm long, lax; flowers white, appearing short-pedicellate, each with a bract to 5 cm long, to 0.5 cm broad; petals slightly longer than the sepals, 8–17 mm long, to 10 mm broad; lip 3-lobed, 15–30 mm long, each lobe fringed; spur 2–6 cm long.

COMMON NAME: White Fringed Orchid.
HABITAT: Swampy areas.
RANGE: Ontario to North Dakota, south to Louisiana, Ohio, and New York.
ILLINOIS DISTRIBUTION: Not common; scattered in the northern two-thirds of the state.
The lovely white flowers appear from mid-June to late July.

12. Habenaria peramoena Gray, Am. Journ. Sci. 38:310. 1840. *Fig. 109.*

Platanthera peramoena (Gray) Gray, Man. Bot. 473. 1848.
Blephariglottis peramoena (Gray) Rydb. in Britt. Man. Fl. N. U. S. 297. 1901.

Rootstocks tuberous; leaves 2–4, cauline, lanceolate, glabrous, to 20 cm long, to 6 cm broad; stem to 80 cm tall, glabrous; inflorescence 6–18 cm long, rather crowded; flowers reddish-purple; sepals and petals about equal, 6–10 mm long, about half as broad; lip 3-lobed, 13–22 mm long, each lobe deeply erose, the terminal lobe with a deep notch; spur 22–30 mm long.

108. *Habenaria leucophaea* (White Fringed Orchid). *a*. Upper part of plant, X¼. *b*. Flower, X2.

109. *Habenaria peramoena* (Purple Fringeless Orchid). *a.* Inflorescence and leaves, X¼. *b.* Flower, X1. *c.* Fruiting branch, X¼.

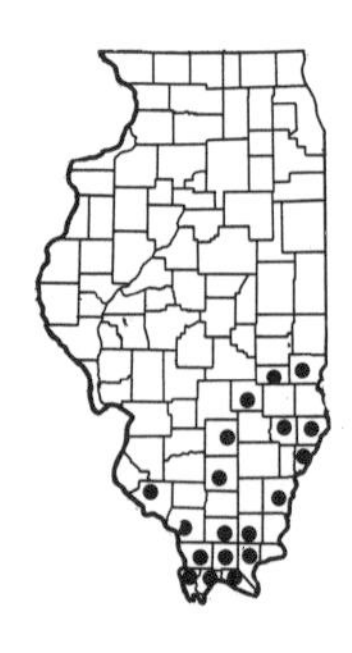

COMMON NAME: Purple Fringeless Orchid.
HABITAT: Low woodlands; wet prairies.
RANGE: New Jersey to Missouri, south to Alabama, Tennessee, and South Carolina.
ILLINOIS DISTRIBUTION: Occasional; restricted to the southern two-fifths of the state.

This is the most common species of *Habenaria* in Illinois. It differs from the similarly colored *H. psycodes* by its erose, notched lip 13–22 mm long.

Flowers are produced from June to early September. This is the most abundant *Habenaria* to occur in the southern one-fourth of Illinois.

13. **Habenaria psycodes** (L.) Spreng. Syst. Veg. 3:693. 1826. *Fig. 110.*

Orchis psycodes L. Sp. Pl. 943. 1753.
Platanthera psycodes (L.) Lindl. Gen. & Sp. Orch. 294. 1835.
Blephariglottis psycodes (L.) Rydb. in Britt. Man. Fl. N. U. S. 296. 1901.

Rootstocks tuberous; leaves (1–) 2–5, cauline, elliptic, glabrous, to 22 cm long, to 5 (–7) cm broad; stem to 75 cm tall, glabrous; inflorescence 5–20 cm long, rather crowded; flowers reddish-purple, each subtended by a bract to 5 cm long, to 0.5 cm broad; sepals and petals about equal, 4–8 mm long, to 5 mm wide; lip 3-lobed, 9–13 mm long, each lobe long-toothed or fringed, the terminal lobe without a notch; spur 16–25 mm long.

COMMON NAME: Purple Fringed Orchid.
HABITAT: Low woodlands; bogs.
RANGE: Newfoundland to Ontario, south to Iowa, Tennessee, and Georgia.
ILLINOIS DISTRIBUTION: Rare; known only from 3 extremely northern counties.

The flowers appear from late June to late August.

110. *Habenaria psycodes* (Purple Fringed Orchid). *a.* Upper part of plant, X¼. *b.* Flower, X1½. *c.* Fruiting branch, X¼.

5. *Liparis* RICH. – Twayblade Orchid

Perennials from bulbs; foliage leaves 2, nearly basal, with several scale leaves present; inflorescence racemose, bracteate; flowers not fragrant; sepals spreading; petals twisted, threadlike; lip entire; anther 1, easily detached; fruit a capsule.

KEY TO THE SPECIES OF Liparis IN ILLINOIS

1. Pedicels 5–10 mm long, as long as the flowers; sepals greenish-white, 10–12 mm long; lateral petals 10–12 mm long, madder-purple; lip 8–12 mm long, madder-purple, flat______1. *L. liliifolia*
1. Pedicels 4–5 mm long, usually a little shorter than the flowers; sepals yellow-green, 4–6 mm long; lateral petals 4–6 mm long, yellowish; lip 4–6 mm long, yellow-green, turned up slightly along the margins________________________________2. *L. loeselii*

1. Liparis liliifolia (L.) Rich. ex Lindl. Bot. Reg. 11:t. 882. 1825. *Fig. 111.*

Ophrys lilifolia L. Sp. Pl. 946. 1753.

Bulb solid, rather thick; leaves 2, nearly basal, broadly elliptic to oval, glabrous, to 15 cm long, 5–10 cm wide, shining; scape to 25 cm tall, angled, glabrous; raceme 5- to 35-flowered; flowers pedicellate, the pedicels 5–10 mm long, as long as the flowers; sepals narrowly lanceolate, twisted, appearing convolute, 10–12 mm long, 2.0–2.5 mm broad, greenish-white; lateral petals linear, threadlike, twisted, madder-purple, 10–12 mm long; lip entire, obovate, apiculate, flat, madder-purple, 8–12 mm long; capsule to 1.5 cm long.

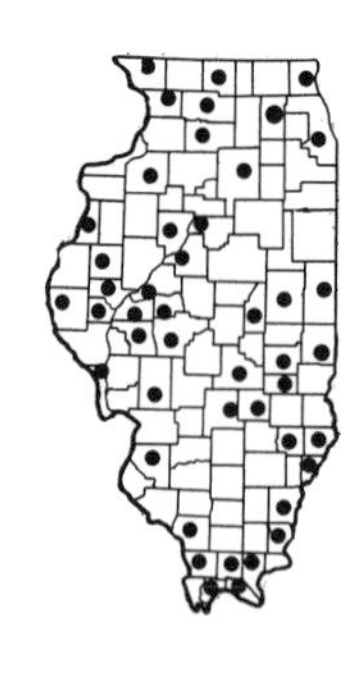

COMMON NAME: Twayblade Orchid.

HABITAT: Rich woodlands.

RANGE: Maine to Minnesota, south to Alabama and Georgia.

ILLINOIS DISTRIBUTION: Occasional; scattered throughout the state.

This species is bigger in all respects than the following species. The flowers may appear the last week in May and continue until mid-July. Although not common, this orchid is one of the more plentiful orchids in Illinois.

111. *Liparis liliifolia* (Twayblade Orchid). *a.* Habit, in flower, X¼. *b.* Flower, X1. *c.* Lip, X1½. *d.* Habit, in fruit, X¼.

2. Liparis loeselii (L.) Rich. Mem. Mus. Par. 4:60. 1817. *Fig. 112.*

Ophrys loeselii L. Sp. Pl. 947. 1753.

Bulb solid, smaller than in the preceding species; leaves 2, nearly basal, lanceolate to lance-ovate, to 15 cm long, 2–3 cm wide, glabrous, shining; scape to 25 cm tall, somewhat angled, glabrous; raceme 3- to 20-flowered; flowers pedicellate, the pedicels 4–5 mm long, usually a little shorter than the flowers; sepals narrowly oblong, 4–6 mm long, about 2 mm broad, yellow-green; lateral petals linear, twisted, 4–6 mm long, yellow-green; lip oblong, entire, apiculate, turned up slightly along the margin, yellow-green, 4–6 mm long; capsule up to 1 cm long; $2n = 26$ (Gadella & Kliphuis, 1963).

COMMON NAME: Lesser Twayblade Orchid.

HABITAT: Low woodlands; bogs.

RANGE: Quebec to Saskatchewan, south to Nebraska, Illinois, Alabama, and North Carolina; Washington.

ILLINOIS DISTRIBUTION: Rare; recorded from seven counties in the northern half of the state.

The flowers are borne in June and July.

6. *Malaxis* sw. – Adder's Mouth Orchid

Perennials from bulbs; leaf 1 (in Illinois species), rarely 2, cauline; inflorescence racemose, bracteate; flowers not fragrant; sepals and the much smaller petals spreading; lip auriculate at base, entire or lobed; anther 1, easily detached; fruit a capsule.

KEY TO THE SPECIES OF Malaxis IN ILLINOIS

1. Pedicels 4–9 mm long; lip bilobed, ascending; upper sepal 1.2–1.6 mm long .... 1. *M. unifolia*
1. Pedicels 1–3 mm long; lip entire, deflexed; upper sepal 2.0–2.5 mm long .... 2. *M. monophylla*

1. Malaxis unifolia Michx. Fl. Bor. Am. 2:157. 1803. *Fig. 113.*

Malaxis ophioglossoides Muhl. ex Willd. Sp. Pl. 4:90. 1805.
Microstylis ophioglossoides (Muhl.) Nutt. Gen. 2:196. 1818.
Microstylis unifolia (Michx.) BSP. Prel. Cat. N. Y. 51. 1888.

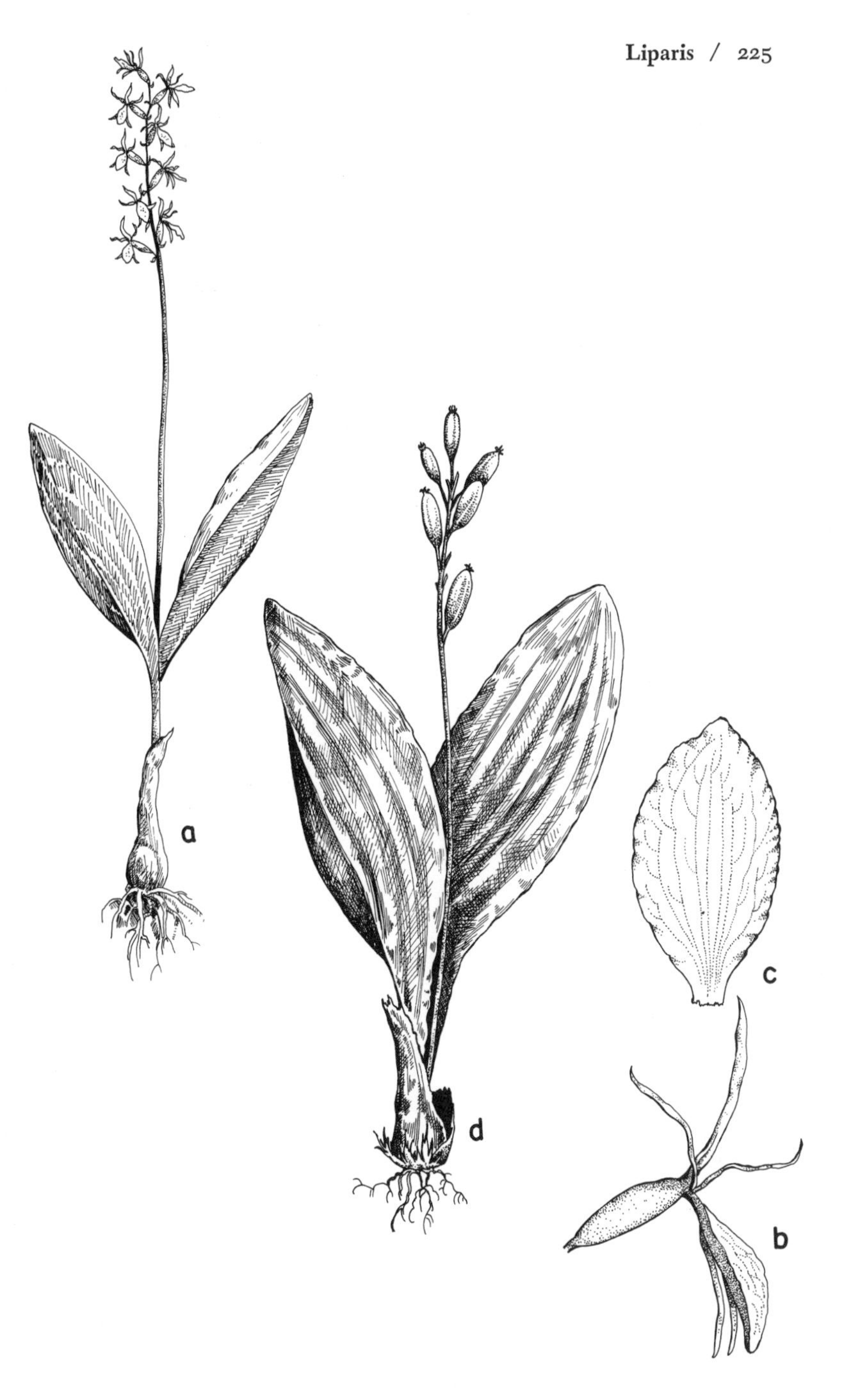

112. *Liparis loeselii* (Lesser Twayblade Orchid). *a.* Habit, in flower, X¼. *b.* Flower, X2½. *c.* Lip, X3¾. *d.* Habit, in fruit, X¼.

113. *Malaxis unifolia* (Adder's Mouth Orchid). *a.* Habit, in flower, X¼. *b.* Inflorescence, X¾. *c.* Flower, X3. *d.* Lip, X4½. *e.* Habit, in fruit, X¼. *f.* Capsule, X1½.

Bulb solid, ovoid; leaf 1 (rarely 2), borne about half-way up the stem, elliptic or broadly elliptic, to 6 cm long, 2–3 cm wide; stem to 25 cm tall; raceme 1.5–2.5 cm thick, 5- to 35-flowered; flowers pedicellate, the pedicels 4–9 mm long; sepals elliptic, obtuse, the upper 1.2–3.0 mm long; petals narrowly linear, about 1 mm long; lip oblong or broadly so, bilobed at the apex with a small tooth between the lobes, 2.0–3.5 mm long, ascending, greenish.

COMMON NAME: Adder's Mouth Orchid.
HABITAT: Dry or moist woodlands.
RANGE: Newfoundland to Saskatchewan, south to Texas and Florida.
ILLINOIS DISTRIBUTION: Rare; known from three west-central counties and one northeastern county.

This and the following are among the smallest orchids in the state. Most specimens have but a single foliage leaf, although rarely a second leaf may be produced. The tiny, delicate flowers open from late June to early August.

2. **Malaxis monophylla** (L.) Sw. var. **brachypoda** (Gray) F. Morris in Morris & Eames, Our Wild Orchids 358. 1929. *Fig. 114.*

Microstylis brachypoda Gray, Ann. Lyc. N.Y. 3:228. 1828.
Malaxis brachypoda (Gray) Fern. Rhodora 28:176. 1926.

Bulb solid, ovoid; leaf 1 (rarely 2), borne below the middle of the stem, broadly elliptic, to 6 cm long, 3–4 cm wide; stem to 25 cm tall; raceme slender, less than 1 cm thick, 4- to 25-flowered; flowers pedicellate, the pedicels 1–3 mm long; sepals broadly lanceolate, subacute, the upper 2.0–2.5 mm long; petals oblanceolate, about 1 mm long; lip broadly cordate at base, long-tapering and pointed at apex, unlobed, 2–3 mm long, deflexed, greenish-white.

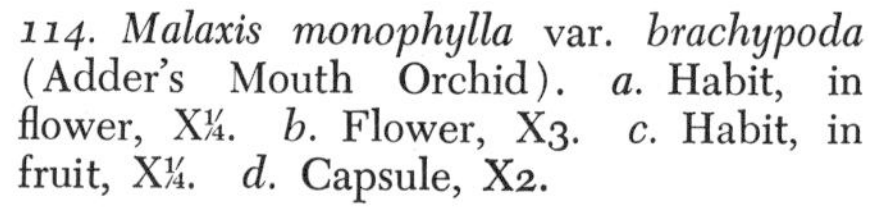

114. Malaxis monophylla var. *brachypoda* (Adder's Mouth Orchid). *a.* Habit, in flower, X¼. *b.* Flower, X3. *c.* Habit, in fruit, X¼. *d.* Capsule, X2.

COMMON NAME: Adder's Mouth Orchid.

HABITAT: Bogs.

RANGE: Labrador to Manitoba, south to Minnesota, Illinois, and Pennsylvania.

ILLINOIS DISTRIBUTION: Very rare; the only collection is by Vasey, and merely states "Bogs, Ill." It is deposited in the herbarium of the Missouri Botanical Garden.

This species is distinguished from the preceding by the shorter pedicels and the unlobed lip. The flowers appear in July. The collection reported here, if actually from Illinois, is considerably south of all other known stations.

7. *Spiranthes* RICH. – Ladies' Tresses

Perennials from tuberous thickened roots; leaves several, often all basal, the upper strongly reduced or seldom like the basal, present or absent at flowering time; inflorescence spicate, bracteate; flowers usually not fragrant, white or yellowish; lateral sepals free, the upper united with the petals; lip clawed, usually erose or crisped, with 2 callosities near the base; anther 1, easily detached; fruit a capsule.

KEY TO THE SPECIES OF Spiranthes IN ILLINOIS

1. Flowers produced in 2–3 rows, the spikes usually crowded; rachis pubescent (see also S. *vernalis* under second number one).
   2. Foliage leaves all basal; lip yellow; flowers produced from May to July ---------------------------------- 1. S. *lucida*
   2. Foliage leaves basal and cauline (at least near base of stem); lip white; flowers produced from August to October.
      3. Sepals and petals 4–5 mm long; lip 4–5 mm long, with two basal, conspicuous, incurved callosities -------- 2. S. *ovalis*
      3. Sepals and petals 7–12 mm long; lip 7–12 mm long, with two basal, rather obscure, rounded callosities ------------ -- 3. S. *cernua*
1. Flowers produced in a single, twisted row, often appearing secund, the spikes laxly-flowered; rachis glabrous (pubescent in S. *vernalis*).
   4. Rachis pubescent; sepals, petals, and the lip 6–10 mm long, the lip yellowish; lowest cauline leaves usually resembling the basal leaves, present at flowering time ---------- 4. S. *vernalis*
   4. Rachis glabrous; sepals, petals, and the lip 3–4 mm long, the lip white or white with a green area; all leaves basal, withered or absent at flowering time.
      5. Tuberous roots several; lip white with a green central area; largest leaves usually 2–4 cm long.
         6. Leaves present but withered at flowering time, thin; spike little spiraling; summit of lip green with a broad white border; plants flowering from mid-June to early September, averaging (according to Fernald [1950]) August 5 ------------------------------------- 5. S. *lacera*
         6. Leaves absent at flowering time, thick; spike strongly spiraling; summit of lip green with a narrow white border; plants flowering from late July to early October,

averaging (according to Fernald [1950]) September 2 ------------------------------------6. *S. gracilis*

5. Tuberous root 1 or occasionally 2 or 3; lip white throughout; largest leaves at most 2 cm long------------7. *S. tuberosa*

1. Spiranthes lucida (H. H. Eaton) Ames, Orchidac. 2:258. 1908. *Fig. 115.*

Neottia plantaginea Raf. Am. Monthly Mag. 2:206. 1818, non *S. plantaginea* Lindl. (1840).

Neottia lucida H. H. Eaton, Trans. Journ. Med. 5:107. 1832.

Spiranthes latifolia Torr. ex Lindl. Gen. & Sp. Orch. pl. 467. 1840.

Ibidium plantagineum (Raf.) House, Bull. Torrey Club 32:381. 1905.

Roots thickened, several; foliage leaves all basal, present at flowering time, oblong to oblanceolate, glabrous, to 15 cm long, 1.5–2.5 cm broad, the cauline leaves all greatly reduced; stem to 25 cm tall, slender; spike 2–8 cm long, crowded, the flowers borne in three ranks, the rachis puberulent; sepals and petals broadly linear, 5–7 mm long; lip yellow, oblong, 5–7 mm long, crisped, with two inconspicuous, rounded callosities at the base.

COMMON NAME: Yellow-lipped Ladies' Tresses.

HABITAT: Wet situations.

RANGE: Maine to Wisconsin, south to Missouri and Virginia.

ILLINOIS DISTRIBUTION: Rare; known only from two northern counties, and not collected since 1897.

The yellow lip distinguishes this species from all other *Spiranthes* in Illinois except *S. vernalis*, a later-flowering species with flowers borne in a single rank.

The flowers of *S. lucida* appear from mid-May to mid-June.

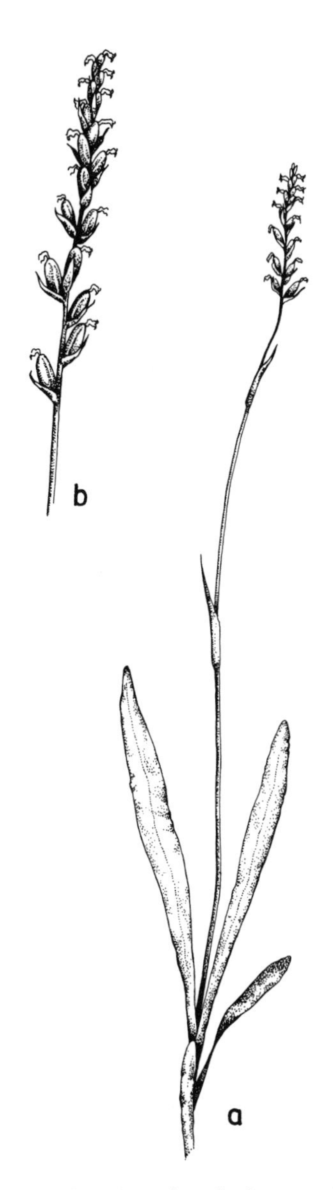

115. *Spiranthes lucida* (Yellow-lipped Ladies' Tresses). *a.* Habit, X⅛. *b.* Inflorescence, X1½.

116. *Spiranthes ovalis* (Ladies' Tresses). *a.* Habit, X¼. *b.* Flower, X2½.

2. **Spiranthes ovalis** Lindl. Gen. & Sp. Orch. 466. 1840. *Fig. 116.*

Ibidium ovale (Lindl.) House, Muhlenbergia 1:128. 1906.

Roots thickened, several; foliage leaves basal and cauline, with only the uppermost cauline leaves greatly reduced, the foliage leaves oblanceolate, to 15 cm long, 1.0–1.5 cm broad, present at flowering time; stem to 35 cm tall, rather slender; spike 2–8 cm long, crowded, the flowers borne in three ranks, the rachis puberulent, the bracts to 1.5 cm long; sepals and petals mostly lanceolate, 4–5 mm long; lip white, ovate, 4–5 mm long, undulate, with two conspicuous, incurved callosities at the base.

COMMON NAME: Ladies' Tresses.

HABITAT: Rich woodlands.

RANGE: Virginia to Missouri, south to Texas and Florida.

ILLINOIS DISTRIBUTION: Rare; known from Jackson, Pulaski, and Union counties in the southern tip of the state, and from Sangamon County in the central part of the state. The first collection was by *H. S. Pepoon and E. G. Barrett* in 1931 from Pulaski County.

This southern species flowers during September and October.

3. **Spiranthes cernua** (L.) Rich. Orch. Eur. Annot. 37. 1817. *Fig. 117.*

Ophrys cernua L. Sp. Pl. 946. 1753.

Gyrostachys ochroleuca Rydb. in Britt. Man. Fl. N. States 300. 1901.

Spiranthes cernua var. *ochroleuca* (Rydb.) Ames, Orchidac. 1:145. 1905.

Ibidium cernuum (L.) House, Bull. Torrey Club 32:381. 1905.

Roots thickened, several; foliage leaves basal and cauline, oblanceolate, to 25 cm long, to 1.5 cm wide, present at flowering time; stem rarely over 50 cm tall, rather stout; spike 3–12 cm long, crowded, the flowers in 2–3 ranks, the rachis puberulent, the bracts to 1.5 cm long; sepals and petals more or less deltoid, 7–12 mm long; lip white, oblong to ovate, 7–12 mm long, crisped, with two rather inconspicuous, rounded basal callosities.

117. Spiranthes cernua (Nodding Ladies' Tresses). *a.* Inflorescence and bracts, X¼. *b.* Flower, X1½. *c.* Inflorescence and leaves, X6.

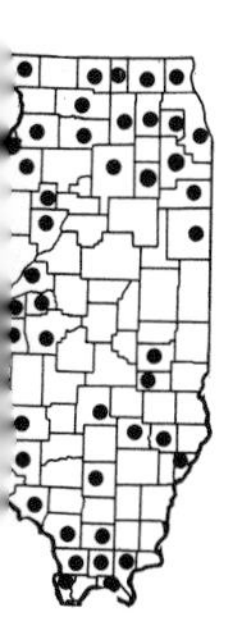

COMMON NAME: Nodding Ladies' Tresses.
HABITAT: Moist or dry situations in usually open areas.
RANGE: Nova Scotia to South Dakota, south to Texas and Florida.
ILLINOIS DISTRIBUTION: Occasional; scattered throughout the state.

There seems to be little reason for maintaining var. *ochroleuca* as distinct. It is said to differ in its greenish-white to stramineous flowers, longer floral bracts, broader lip, and drier habitats, but these do not seem to hold for Illinois material.

This is the most common *Spiranthes* in Illinois and one of the commonest orchids in the state. The flowers appear from early August to early October.

4. **Spiranthes vernalis** Engelm. & Gray, Boston Journ. Nat. Hist. 5:236. 1845. *Fig. 118.*

Ibidium vernale (Engelm. & Gray) House, Bull. Torrey Club 32:381. 1905.

Roots thickened, several; upper cauline leaves bladeless, the lower and the basal leaves linear to linear-lanceolate, to 25 cm long, 1.5–2.5 cm broad, present at flowering time; stem to 60 cm tall, slender; spike 5–15 cm long, laxly flowered, 1-ranked, the bracts to 1.5 cm long; rachis pubescent; sepals and petals linear-lanceolate, 6–10 mm long; lip yellowish, ovate, to 8 mm long, to 6 mm broad, crisped, with two conspicuous, incurved callosities at the base.

COMMON NAME: Ladies' Tresses.
HABITAT: Rich woods; prairies.
RANGE: Massachusetts to Kansas, south to Texas and Florida.
ILLINOIS DISTRIBUTION: Rare; first collected in 1865 by Hall from Menard County; subsequently collected in St. Clair, Effingham, Massac, and Pope counties.

This is the only species of *Spiranthes* in Illinois which has a pubescent rachis and flowers borne in a single rank. The flowers appear in July and August.

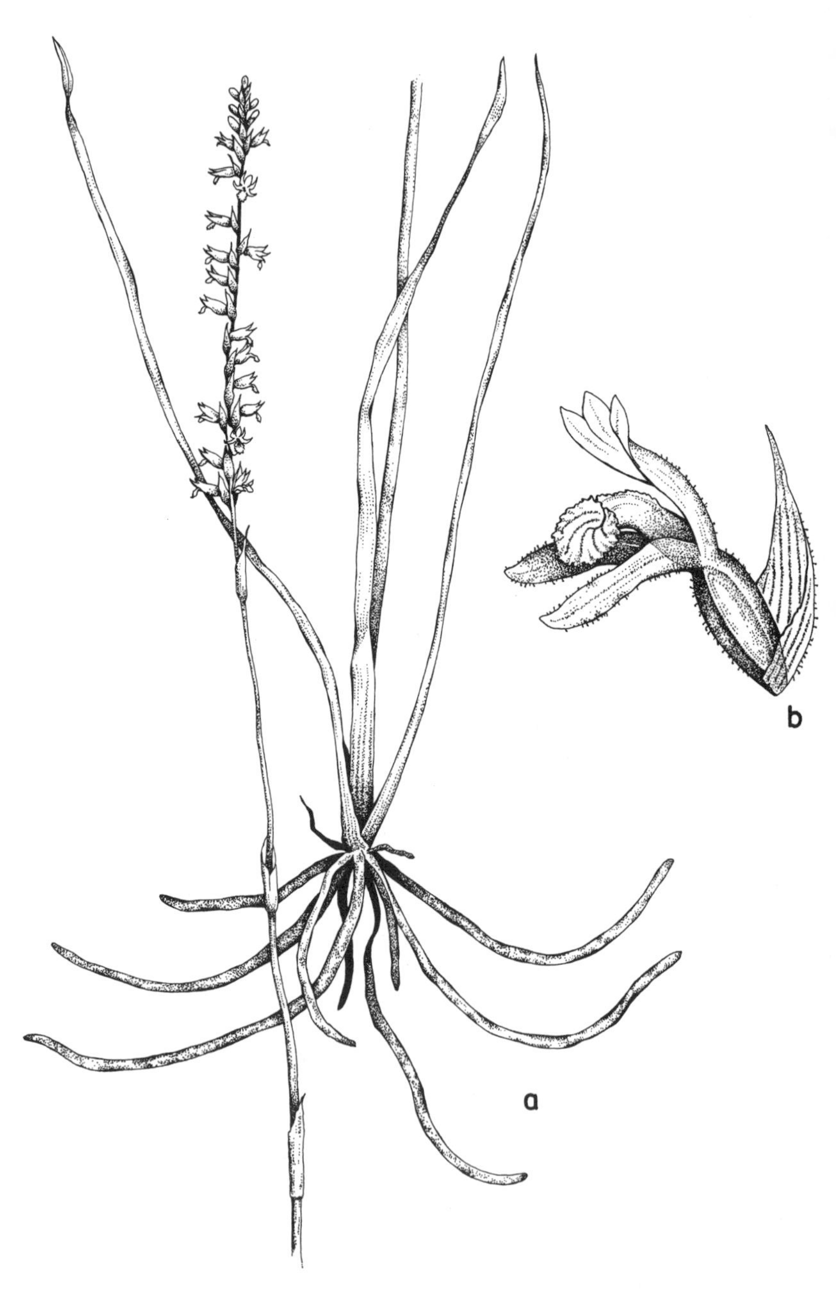

118. Spiranthes vernalis (Ladies' Tresses). *a.* Inflorescence and base of plant, X¼. *b.* Flower, X1½.

5. **Spiranthes lacera** (Raf.) Raf. Herb. Raf. 44. 1833. *Fig. 119.*

Neottia lacera Raf. Fl. Ludov. 171. 1817.

Roots thickened, several; foliage leaves all basal, thin, oval to oblong, to 5 cm long, to 2 cm broad, present but usually withering at flowering time, the cauline leaves bladeless; stem to 50 cm tall, very slender; spike 2–12 cm long, laxly flowered, the flowers in a single, scarcely twisted rank; flowers scarcely open; rachis glabrous; sepals and petals linear, about 4 mm long; lip white with a green central area, with a broad white margin at the summit, oblong, about 4 mm long, crisped, minutely pubescent above.

COMMON NAME: Slender Ladies' Tresses.

HABITAT: Mostly dry woodlands.

RANGE: Quebec to Mackenzie, south to Illinois and North Carolina.

ILLINOIS DISTRIBUTION: Rare; confined to northern Illinois.

This species flowers from mid-June to early September. After studying several hundred specimens of "S. *gracilis*" from all of eastern North America, I must conclude with Fernald that we are dealing with two distinct entities, best treated as species. The differences enumerated in the key between S. *lacera* and S. *gracilis* seem to be constant and are correlated with flowering time and species range.

6. **Spiranthes gracilis** (Bigel.) Beck, Bot. 343. 1833. *Fig. 120.*

Neottia gracilis Bigel. Fl. Bost. 322. 1824.

Ibidium gracile (Bigel.) House, Bull. Torrey Club 32:381. 1905.

Roots thickened, several; foliage leaves all basal, thick, the veining obscure unless held up to the light, oval to oblong, to 6 cm long, to 2 cm broad, absent at flowering time, the cauline leaves bladeless; stem to 60 cm tall, slender; spike 4–15 cm long, laxly flowered, the flowers in a single, strongly twisted rank; flowers open; rachis glabrous; sepals and petals linear, about 4 mm long; lip white with a green central area, with a narrow white margin at the summit, oblong, about 4 mm long, crisped, minutely pubescent above.

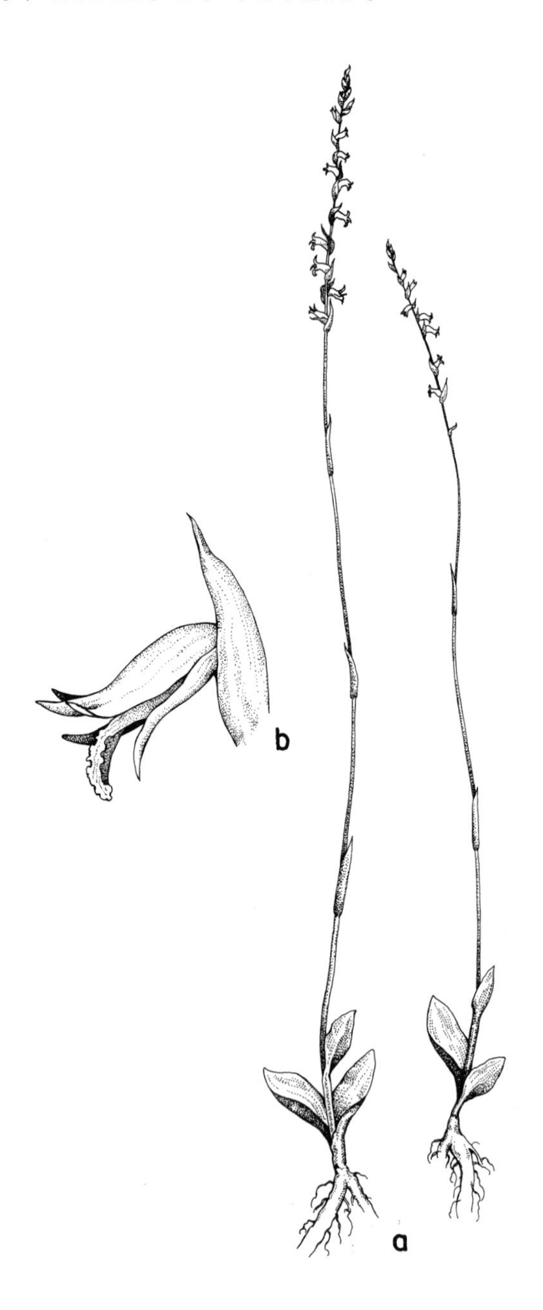

119. *Spiranthes lacera* (Slender Ladies' Tresses). *a*. Habit, X¼. *b*. Flower, X3½.

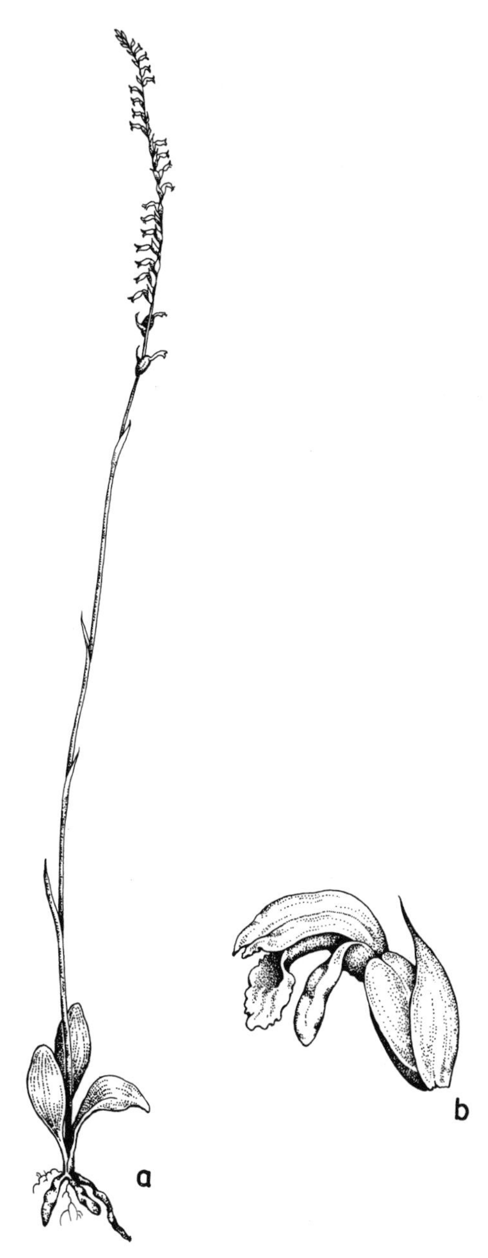

120. *Spiranthes gracilis* (Slender Ladies' Tresses). *a.* Habit, X¼. *b.* Flower, X3½.

COMMON NAME: Slender Ladies' Tresses.
HABITAT: Dry, open woodlands.
RANGE: Maine to Kansas, south to Texas and Florida.
ILLINOIS DISTRIBUTION: Rare; restricted to the southern one-third of the state.

The binomial S. *gracilis* in this work is being restricted to Illinois specimens from the southern one-third of the state. Northern Illinois specimens previously referred to as S. *gracilis* are here considered to be S. *lacera*. Specimens from Randolph and Jackson counties, previously identified as S. *gracilis*, are really S. *tuberosa* with 2–3 roots.

Flowering time is from late July to early October.

7. **Spiranthes tuberosa** Raf. Herb. Raf. 45. 1833. *Fig. 121.*

Spiranthes beckii Lindl. Gen. & Sp. Orch. 472. 1840.
Spiranthes grayi Ames, Rhodora 6:44. 1904.
Spiranthes tuberosa var. *grayi* (Ames) Fern. Rhodora 48:189. 1946.

Root thickened, solitary (or rarely 2–3); foliage leaves all basal, oblong to narrowly ovate, to 2 cm long, 1.0–1.5 cm broad, absent at flowering time, the cauline leaves bladeless; stem to 25 cm tall, very slender; spike 1.5–6.5 cm long, very laxly flowered, the flowers in a single, twisted rank; rachis glabrous; sepals and petals linear-lanceolate, 2–4 mm long; lip white throughout, broadly oblong, 3.0–3.5 mm long, erose or crisped, densely puberulent above.

COMMON NAME: Little Ladies' Tresses.
HABITAT: Dry, open woodlands.
RANGE: Massachusetts to Missouri, south to Texas and Florida.
ILLINOIS DISTRIBUTION: Not common, confined to the southern one-fourth of the state.

This is the most slender orchid in Illinois. Its tiny flowers are produced from late July to early October. It differs from all other *Spiranthes* in Illinois in the presence of a single thickened root. (Some specimens are occasionally found with 2–3 roots; these are often confused with S. *gracilis*.) The entirely white lip also distinguishes it from S. *gracilis* or S. *lacera*.

121. Spiranthes tuberosa (Little Ladies' Tresses). *a.* Habit, X¼. *b.* Inflorescence and bracts, X¼. *c.* Inflorescence, X1¾. *d.* Flower, X3½.

Although some botanists have maintained that our specimens should be called a variety of S. *tuberosa* (as var. *grayi*) or even a separate species (S. *grayi*), the only difference is in degree of spiraling of the inflorescence. A study of many specimens from throughout the entire range of S. *tuberosa* reveals that the spiraling character should not be relied upon.

8. *Goodyera* R. BR. – Rattlesnake Plantain Orchid

Perennial from a short, rather fleshy rhizome; leaves all basal, conspicuously veiny, petiolate, evergreen; stem glandular-pubescent; inflorescence a spike-like raceme, bracteate; flowers not fragrant; lateral sepals free, the upper adnate to the lateral petals, forming a galea over the lip; lip sac-like, beaked at the summit; anther 1, easily detached; fruit a capsule.

Only the following species occurs in Illinois.

1. Goodyera pubescens (Willd.) R. Br. in Ait. Hort. Kew. 5:198. 1813. *Fig. 122.*

Neottia pubescens Willd. Sp. Pl. 4:76. 1805.
Peramium pubescens (Willd.) MacM. Met. Minn. 172. 1892.
Epipactis pubescens (Willd.) A. A. Eaton, Proc. Biol. Soc. Wash. 21:65. 1908.

Plants colonial from a fleshy, creeping rhizome; leaves basal, ovate-lanceolate to ovate, to 6 cm long, 2–4 cm wide, dark green with conspicuous white venation; scape to 35 (–40) cm tall; raceme 3–10 cm long, crowded; bracts lance-subulate; sepals and petals glandular-pubescent, broadly lanceolate to ovate, 3.5–5.0 mm long; lip subglobose, white, sac-like at the base, 3.5–4.0 mm long, with a straight beak up to 1 mm long.

COMMON NAME: Rattlesnake Plantain Orchid.

HABITAT: Rich, moist woodlands.

RANGE: Newfoundland to Ontario, south to Missouri, Alabama, and Florida.

ILLINOIS DISTRIBUTION: Rare (although sometimes abundant where found); restricted to the extreme northern and the extreme southern counties; absent elsewhere.

The handsome evergreen rosettes make this an easy species to recognize during the winter. The small white flowers are produced from late June to mid-August.

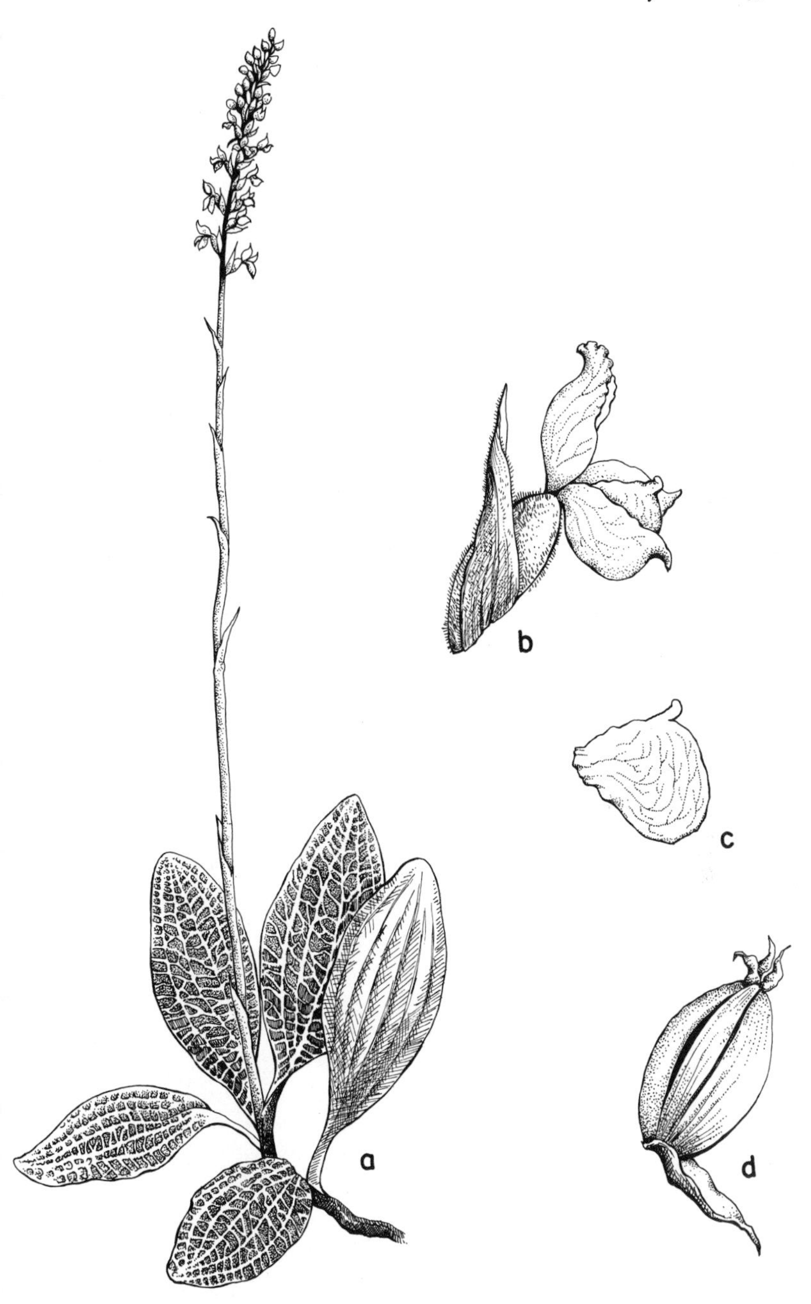

122. *Goodyera pubescens* (Rattlesnake Plantain Orchid). *a*. Habit, X¼. *b*. Flower, X2½. *c*. Lip, X2½. *d*. Capsule, X1.

9. *Pogonia* JUSS. – Pogonia

Perennial from a short rhizome with thickened roots; leaves basal plus one cauline leaf near the base of the stem; inflorescence terminal, 1- to 2-flowered, each flower subtended by a foliaceous bract; flowers fragrant; sepals and petals free, spreading; lip fringed on the margin, bearded on the upper face; anther 1, easily detached; fruit a capsule.

Only the following species occurs in Illinois.

1. **Pogonia ophioglossoides** (L.) Ker, Bot. Reg. 2:pl. 148. 1816. *Fig. 123.*

Arethusa ophioglossoides L. Sp. Pl. 951. 1753.

Rhizome short, with elongated thickened roots; leaves lanceolate to elliptic, the basal petiolate, the single cauline leaf sessile, to 10 cm long, to 2.5 cm broad; stem to 65 cm tall; flowers 1–3, fragrant; bracteal leaf lanceolate, foliaceous, smaller than the single cauline leaf; sepals and petals oblong, obtuse, pink, 15–25 mm long; lip elliptic, 15–20 mm long, pink, with a yellow beard; capsules to 3 cm long.

COMMON NAME: Snake-mouth.

HABITAT: Low ground.

RANGE: Newfoundland to Ontario, south to Texas and Florida.

ILLINOIS DISTRIBUTION: Rare; known from five counties in the northern one-fourth of the State.

This handsome orchid flowers in June and July. The flowers may be various intensities of pink.

10. *Isotria* RAF. – Whorled Pogonia

Perennial with a short rhizome; leaves reduced to sheaths, except for a whorl of involucral leaves beneath the inflorescence; inflorescence composed of a single (rarely two) flower; sepals and petals free; lip three-lobed, with a crest down the middle; fruit a capsule.

Only the following species occurs in Illinois.

123. Pogonia ophioglossoides (Pogonia). *a.* Habit, in flower, X¼. *b.* Flower, X1. *c.* Habit, in fruit, X¼.

1. **Isotria verticillata** (Willd.) Raf. Med. Repos. 5:357. 1808. *Fig. 124.*

Arethusa verticillata Willd. Sp. Pl. 4:81. 1805.

Pogonia verticillata (Willd.) Nutt. Gen. 2:192. 1818.

Rhizome short, with several horizontal, cord-like roots; stem glabrous, purplish, to 35 cm tall, bearing several scaly sheaths near base and a single whorl of 4–8 leaves beneath the inflorescence; involucral leaves obovate-elliptic, subacute at apex, subcuneate at base, to 10 cm long, to 6 cm broad; flower 1, on an erect peduncle shorter than the subtending leaves; sepals linear, to 6 cm long, yellow-green below, purple above; petals up to half as long as the sepals, but slightly broader, greenish-yellow; lip elliptic-obovate, green, 3-lobed, white, except for the purple stripes on the lateral lobes; capsule to 3.5 cm long, on an erect peduncle to 6 cm long.

COMMON NAME: Whorled Pogonia.

HABITAT: Lowland woods (in Illinois).

RANGE: Maine to Michigan, south to Texas and Florida.

ILLINOIS DISTRIBUTION: Known only from two stations at Cove Spring, Pope County, where over 100 plants occur. First collected by John Schwegman in July, 1967.

At the Illinois station, associated species are *Uvularia sessilifolia, Triadenum tubulosum,* and *Tipularia discolor.* Woodland species are white oak, sugar maple, and pignut hickory.

The single whorl of leaves on the stem gives this species a strong resemblance to *Medeola virginiana.*

The flowers, which are very infrequently formed, are much larger than one would expect for such a small plant. They appear in late May or early June.

11. *Triphora* NUTT. – Nodding Pogonia

Perennial from solid, tuber-like bulbs; leaves cauline, several; flowers axillary, nodding, appearing pedicellate, lasting but a day, not fragrant; sepals and petals free, ascending; lip 3-lobed; anther 1, easily detached; fruit a capsule.

Only the following species occurs in Illinois.

124. *Isotria verticillata* (Whorled Pogonia). *a*. Habit, X¼. *b*. Capsule, X⅔.

1. **Triphora trianthophora** (Sw.) Rydb. in Britt. Man. Fl. N. U. S. 298. 1901. *Fig. 125.*

Arethusa trianthophora Sw. Kongl. Sv. Vet. Acad. Handl. II. 21:230. 1800.

Triphora pendula Nutt. Gen. 2:193. 1818.

Pogonia pendula (Nutt.) Lindl. Bot. Reg. pl. 908. 1825.

Pogonia trianthophora (Sw.) BSP. Prel. Cat. N. Y. 52. 1888.

Bulbous perennial with horizontal stolons; leaves several, cauline, sessile, ovate, glabrous, 1–2 cm long, 0.5–1.5 cm broad; stem to 25 cm tall, glabrous, reddish-brown, very fragile; flowers usually 3, axillary, pedicellate, ascending at first, later pendant; sepals and petals lanceolate, 15–20 mm long, to 6 cm broad, white or pinkish; lip white, 3-lobed, 15–20 mm long, each lobe with a green stripe; capsule to 2.5 cm long.

COMMON NAME: Nodding Pogonia.

HABITAT: Rich woodlands.

RANGE: Maine to Iowa, south to Texas and Florida.

ILLINOIS DISTRIBUTION: Not common; scattered throughout the state.

This delicate orchid is one of the last to flower, bringing forth its blossoms from mid-August to mid-October. It may be found where the humus is exceptionally thick.

12. *Epipactis* sw. – Helleborine

Perennial from a thickened rootstock; leaves several, cauline, clasping the stem; inflorescence terminal, racemose, long-bracteate; flowers large, not fragrant; sepals and petals free, keeled; lip sac-like at the base; anther 1, easily detached; fruit a capsule.

Only the following species occurs in Illinois.

1. **Epipactis helleborine** (L.) Crantz, Stirp. Austr. ed. 2, 6:467. 1769. *Fig. 126.*

Serapias helleborine L. Sp. Pl. 949. 1753.

Serapias latifolia Huds. Fl. Angl. ed. I. 341. 1762.

Epipactis latifolia All. F. Pedem. 2:151. 1785.

125. *Triphora trianthophora* (Nodding Pogonia). *a.* Habit, in flower, X⅙. *b.* Flower, X1. *c.* Lip, X2. *d.* Habit, in fruit, X⅙.

Leaves several, cauline, lanceolate to ovate, to 15 cm long, to 8 cm broad; stem to 1 m tall, short-pubescent below, glabrous above; inflorescence racemose, 5–25 cm long, 5- to 25-flowered; flowers pedicellate, each subtended by a shorter or longer linear-lanceolate bract; sepals and petals broadly lanceolate, 10–15 mm long, 4–5 mm broad, green suffused with purple; lip sac-like, green suffused with purple, to 10 mm long, to 5.5 mm broad, with two tubercles at the base; capsule pendent, to 1 cm long; $2n = 38$ (Gadella & Kliphuis, 1963).

COMMON NAME: Helleborine.

HABITAT: Disturbed woodlands.

RANGE: Native of Europe; naturalized in various parts of the United States and Canada.

ILLINOIS DISTRIBUTION: Rare, but becoming more frequent in Cook, DuPage, and Lake counties. The first Illinois collection was made by J. A. Steyermark in 1954.

This is the only orchid in the Illinois flora which is not native. It appears to be becoming rather widespread in the United States. The flowers appear in July and August. Basal leaves are bladeless; the largest leaves occur midway on the stem.

13. *Corallorhiza* CHAT. – Coral-root Orchid

Yellow, brown, or purple saprophytic perennials from a series of coral-like roots; leaves all reduced to non-green, sheathing scales; inflorescence terminal, racemose; sepals and petals free, 1- to 3-nerved, spreading or arched-ascending over the lip; lip entire or shallowly lobed, with longitudinal ridges on the upper face; anther 1, easily detached; fruit a capsule.

KEY TO THE SPECIES OF Corallorhiza IN ILLINOIS

1. Lip with two lateral lobes or teeth; plants flowering generally from mid-May to mid-August.
   2. Stems yellowish; sepals and petals yellow-green, rarely spotted with purple, 4–5 mm long, 1-nerved; spur absent; lip 4–5 mm long, with 2 short lateral lobes; capsule greenish, 6–10 mm long; rhizome white---------------------------------1. *C. trifida*
   2. Stems purplish; sepals and petals white with purple spots, 6–8 mm long, 3-nerved; spur present; lip 6–8 mm long, with 2

126. *Epipactis helleborine* (Helleborine). *a.* Habit, in flower, X¼. *b.* Flower, X1.

well-developed lateral teeth or lobes; capsule brownish, 10–20 mm long; rhizome brown ---------------------- 2. *C. maculata*

1. Lip entire or erose, not lobed or toothed; plants flowering generally from March to mid-May or from mid-August to October.
   3. Stems purple; sepals and petals greenish-yellow, spotted with purple, linear-lanceolate, 6–8 mm long; lip 5–6 mm long, notched at apex, long-clawed; capsule 8–12 mm long; flowering in early spring -------------------------- 3. *C. wisteriana*
   3. Stems purple or brown below, greenish above; sepals and petals purplish-green to purple, oblong, 3–5 mm long; lip 3–4 mm long, undulate at apex but not notched, short-clawed or clawless; capsule 5–8 mm long; flowering in autumn -------- -- 4. *C. odontorhiza*

1. **Corallorhiza trifida** Chat. Spec. Inaug. Coral. 8. 1760. *Fig. 127.*

Corallorhiza innata R. Br. in Ait. Hort. Kew. 5:209. 1813.

Corallorhiza trifida var. *verna* (Nutt.) Fern. Rhodora 48:196. 1946.

Rhizome coralloid, white; leaves reduced to tubular, clasping sheaths; stem yellowish, to 25 cm tall; raceme 2.0–7.5 cm long, 2- to 12-flowered; sepals and petals linear-lanceolate, 1-nerved, 4–5 mm long, yellow-green, rarely spotted with purple; spur absent; lip white, with a few purple spots, 4–5 mm long, with 2 short lateral lobes; capsule greenish, 6–10 mm long; 2n = 42 (Löve & Löve, 1948).

COMMON NAME: Pale Coral-root Orchid.

HABITAT: Moist woodlands.

RANGE: Labrador to Alaska, south to Oregon, Illinois, Tennessee, and Georgia; Eurasia.

ILLINOIS DISTRIBUTION: Very rare; the only specimen seen was from St. Clair County, collected before 1900.

There seems to be no reason to recognize our specimen as var. *verna,* a taxon with smaller perianth than typical var. *trifida.*

This species is similar to *C. maculata,* but begins to flower about a month earlier. The only collection from Illinois was collected in mid-May while the plant was in flower.

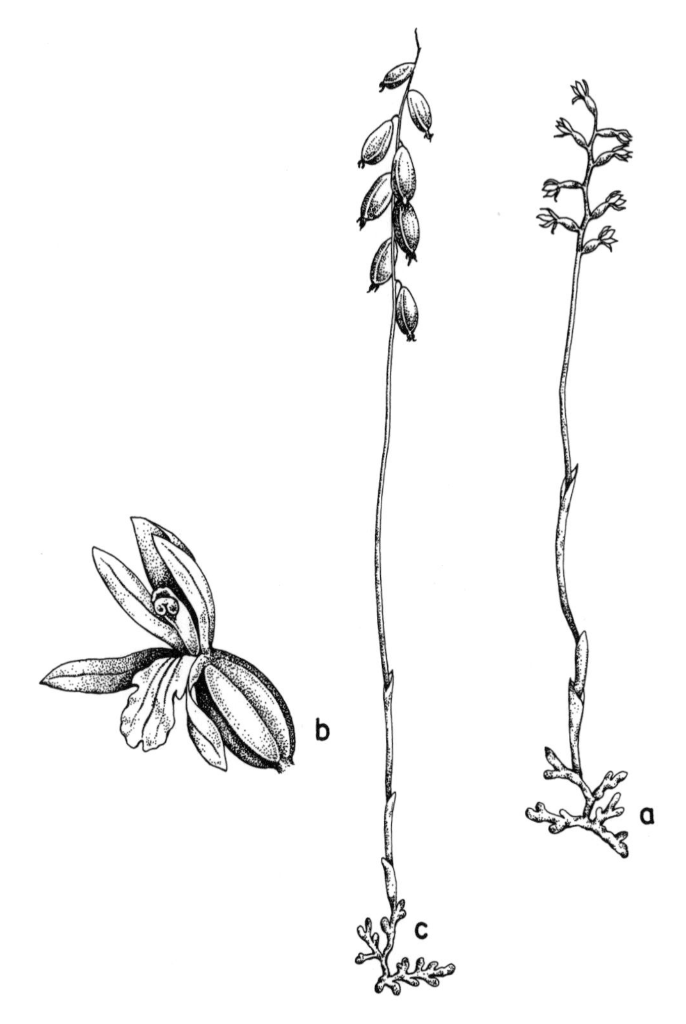

127. Corallorhiza trifida (Pale Coral-root Orchid). *a.* Habit, in flower, ¼. *b.* Flower, X2½. *c.* Habit, in fruit, X¼.

2. Corallorhiza maculata Raf. Am. Month. Mag. 2:119. 1817. *Fig. 128.*

Corallorhiza multiflora Nutt. Journ. Acad. Phila. 3:138. 1823. Rhizome coralloid, brown; leaves reduced to tubular clasping sheaths; stem purplish, to 75 cm tall; raceme 5–20 cm long, 10- to 40-flowered; sepals and petals oblong to oblanceolate, 3-nerved. 6–8 mm long, to 2.5 mm broad, white with purple spots; spur present; lip white, with several purple spots, 6–8 mm long, to 2.5 mm broad, with two conspicuous lateral lobes or teeth; capsule pendent, brownish, 10–25 mm long.

COMMON NAME: Spotted Coral-root Orchid.

HABITAT: Woodlands, particularly those rich in humus.

RANGE: Newfoundland to British Columbia, south to California, Illinois, Tennessee, and North Carolina.

ILLINOIS DISTRIBUTION: Rare; known from six counties in the northern two-fifths of the state.

Variation in the amount of purple-spotting and the color of the stem exists.

The flowers are borne from June to mid-August, while those of *C. trifida* are borne in May. The distinguishing characters given in the key readily separate *C. maculata* from *C. trifida.*

3. **Corallorhiza wisteriana** Conrad, Journ. Acad. Nat. Sci. Phila. 6:145. 1829. *Fig. 129.*

Rhizome coralloid; stem purplish, to 40 cm tall; raceme 3–12 cm long, 10- to 15-flowered; sepals and petals linear-lanceolate, greenish-yellow with purple flecks, 6–10 mm long, to 2 mm broad; lip white, spotted with purple, notched at apex, 5–6 mm long, 4–5 mm broad, long-clawed; capsule 8–12 mm long, pendent.

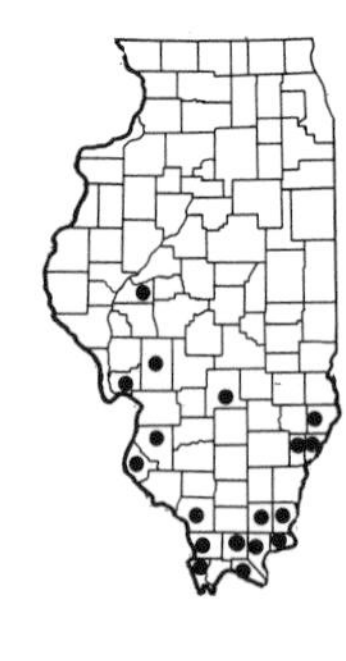

COMMON NAME: Spring Coral-root Orchid; Wister's Coral-root Orchid.

HABITAT: Rich woodlands.

RANGE: New Jersey to South Dakota, south to Texas and Florida.

ILLINOIS DISTRIBUTION: Not common; scattered in the southern half of the state.

This is the first *Corallorhiza* to flower in Illinois, opening its blossoms in mid-March to mid-May.

128. *Corallorhiza maculata* (Spotted Coral-root Orchid). *a.* Habit, in flower, X¼. *b.* Flower, X2. *c.* Habit, in fruit, X¼.

129. *Corallorhiza wisteriana* (Spring Coral-root Orchid). *a.* Inflorescence and upper bracts, X⅝. *b.* Flower, X1½. *c.* Lip, X3. *d.* Habit, in fruit, X⅝.

4. **Corallorhiza odontorhiza** (Willd.) Nutt. Gen. 2:197. 1818. *Fig. 130.*

Cymbidium odontorrhizon Willd. Sp. Pl. 4:110. 1805.

Rhizome coralloid; leaves reduced to tubular clasping sheaths; stem purple or brown below, greenish above, to 30 cm tall; raceme 2–6 cm long, 3- to 15-flowered; sepals and petals oblong, purplish-green to purple, 3–5 mm long; lip orbicular, white, spotted with purple, undulate at apex but not notched, 3–4 mm long, 4–5 mm broad, short-clawed or clawless; capsule pendent, 5–8 mm long.

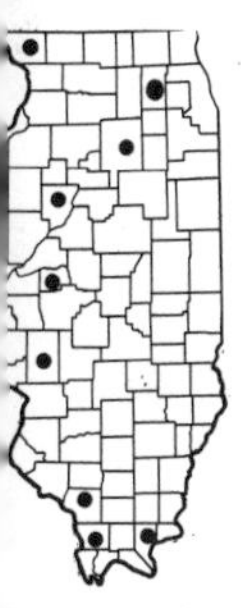

COMMON NAME: Fall Coral-root Orchid; Late Coral-root Orchid.

HABITAT: Woodlands.

RANGE: Maine to Minnesota, south to Texas and Florida.

ILLINOIS DISTRIBUTION: Rare; scattered in the western counties of the state.

This is the last of the coral-roots to flower. The flowers appear from mid-August until October. It is the smallest of the Illinois coral-roots.

14. *Hexalectris* RAF. – Crested Coral-root Orchid

Yellow or brown saprophytic perennial from a cluster of coral-like roots; leaves all reduced to non-green, sheathing scales; inflorescence terminal, racemose; sepals and petals free, 5- to 7-nerved; lip 3-lobed, with several longitudinal ridges on the upper face; anther 1, easily detached; fruit a capsule.

Only the following species occurs in Illinois.

1. **Hexalectris spicata** (Walt.) Barnh. Torreya 4:121. 1904. *Fig. 131.*

Arethusa spicata Walt. Fl. Car. 222. 1788.

Bletia aphylla Nutt. Gen. 2:194. 1818.

Hexalectris aphylla (Nutt.) Raf. Neogenyton 4. 1825.

Rhizome coralloid; stem yellow or brown, with purplish scales, to 60 cm tall; raceme 10–20 (–25) cm long, 5- to 15-flowered; flowers short-pedicellate, each subtended by a deltoid bract; sepals and petals oblanceolate, obtuse to subacute, 17–21 mm long, to 10 mm broad, yellowish, with brownish-purple markings; lip 12–18 mm long, to 15 mm broad, 3-lobed, yellowish

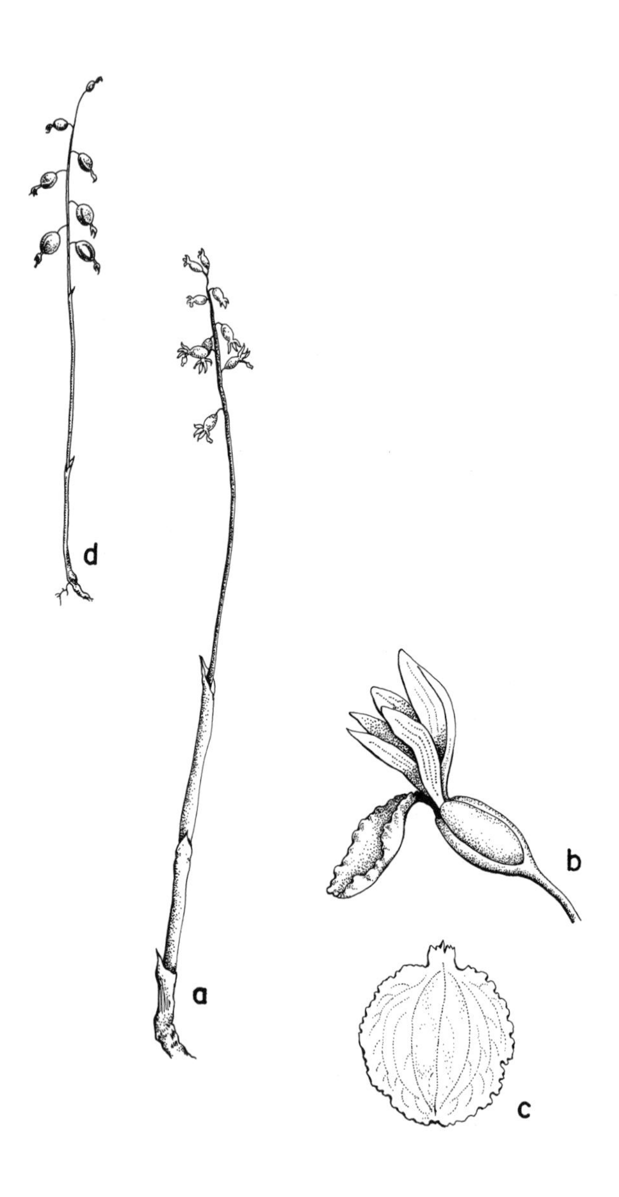

130. *Corallorhiza odontorhiza* (Fall Coral-root Orchid). *a.* Habit, in flower, X¼. *b.* Flower, X2½. *c.* Lip, X3. *d.* Habit, in fruit, X¼.

131. *Hexalectris spicata* (Crested Coral-root Orchid). *a.* Inflorescence and lower part of stem, X⅜. *b.* Flower, X1½. *c.* Fruiting cluster, X⅜.

with purple veins, conspicuously veiny and ridged; capsule pendent, to 2.5 cm long.

COMMON NAME: Crested Coral-root Orchid.
HABITAT: Dry woodlands; limestone ledges.
RANGE: Maryland to Missouri, south to Arizona and Florida; Mexico.
ILLINOIS DISTRIBUTION: Rare; known from four counties in extreme southern Illinois.

This species is closely related to *Corallorhiza*, but is larger in all respects. In addition, the lip is strongly 3-lobed, heavily veiny, and bears several distinct longitudinal ridges. The flowers appear from late June to mid-September. The first Illinois collection was in 1949 by G. S. Winterringer.

15. *Aplectrum* (Nutt.) TORR. – Putty-root Orchid

Perennial from corms; leaf 1, basal, overwintering, usually absent at flowering time; inflorescence racemose, bracteate; sepals and petals free; lip unequally 3-lobed, with 3 conspicuous ridges on the face; anther 1, easily detached; fruit a capsule.

Only the following species comprises the genus.

1. **Aplectrum hyemale** (Muhl.) Torr. Compend. Fl. N. & Mid. States 322. 1826. *Fig. 132.*

Cymbidium hyemale Muhl. ex Willd. Sp. Pl. 4:107. 1805.

Corms globoid, 2–4 in a chain; leaf 1, basal, broadly elliptic, to 15 cm long, to 8 cm broad, blue-green, coarsely veined, usually absent at flowering time; scape to 50 cm tall, bearing tubular sheathing scales; raceme to 15 (–20) cm long, 8- to 17-flowered, bracteate; sepals and petals oblanceolate to oblong, 10–15 mm long, to 5 mm broad, yellowish tinged with madder-purple, the sepals spreading, the petals projecting over the column; lip white, marked with purple, 10–15 mm long; capsule pendent, to 2.5 cm long.

132. *Aplectrum hyemale* (Putty-root Orchid). *a*. Leaf, X¼. *b*. Fruiting branch, X¼. *c*. Inflorescence, X¼. *d*. Flower, X1¼. *e*. Lip, X1¾. *f*. Capsule, X¾.

COMMON NAME: Putty-root Orchid; Adam-and-Eve.
HABITAT: Rich woodlands.
RANGE: Quebec to Saskatchewan, south to Arkansas and Georgia.
ILLINOIS DISTRIBUTION: Occasional; scattered throughout the state, but apparently more common in the extreme southern counties.

The frequently paired corms account for the common name Adam-and-Eve.

A new leaf is produced in the fall and it persists until late spring. It withers as the flower stalk is produced. It differs from the overwintering leaf of *Tipularia discolor* by being green on both sides. Flowers open in May and June.

16. *Tipularia* NUTT. – Crane-fly Orchid

Perennial from rhizomes and tubers; leaf 1, basal, overwintering, usually absent at flowering time; inflorescence racemose, without bracts; sepals and petals free; lip unequally 3-lobed, spurred; anther 1, easily detached; fruit a capsule.

Only the following species occurs in Illinois.

1. **Tipularia discolor** (Pursh) Nutt. Gen. Am. 2:195. 1818. *Fig. 133.*

Orchis discolor Pursh, Fl. Am. Sept. 2:580. 1814.
Limodorum unifolium Muhl. ex Nutt. Gen. Am. 2:195. 1818.
Plectrurus discolor (Pursh) Raf. Neogenyt. 4. 1825.
Tipularia unifolia (Muhl.) BSP. Prel. Cat. 51. 1888.

Tubers several, connected by a slender rhizome; leaf 1, basal, ovate, to 12 cm long, to 8 cm broad, green above, purple beneath, coarsely veined, usually absent at flowering time, long-petiolate; scape to 40 cm tall, bearing sheathing, often purplish scales; raceme to 25 cm long, 10- to 25-flowered, the flowers more or less pendent, bractless; sepals and petals oblong, spreading, 6–8 mm long, to 3 mm broad, greenish-purple with purple veins; lip pale purple, 3-lobed, with two of the lobes very small and basal, 4–7 mm long, 2–3 mm broad, with a slender, curved spur 15–20 mm long; capsule pendent, to 1.5 cm long.

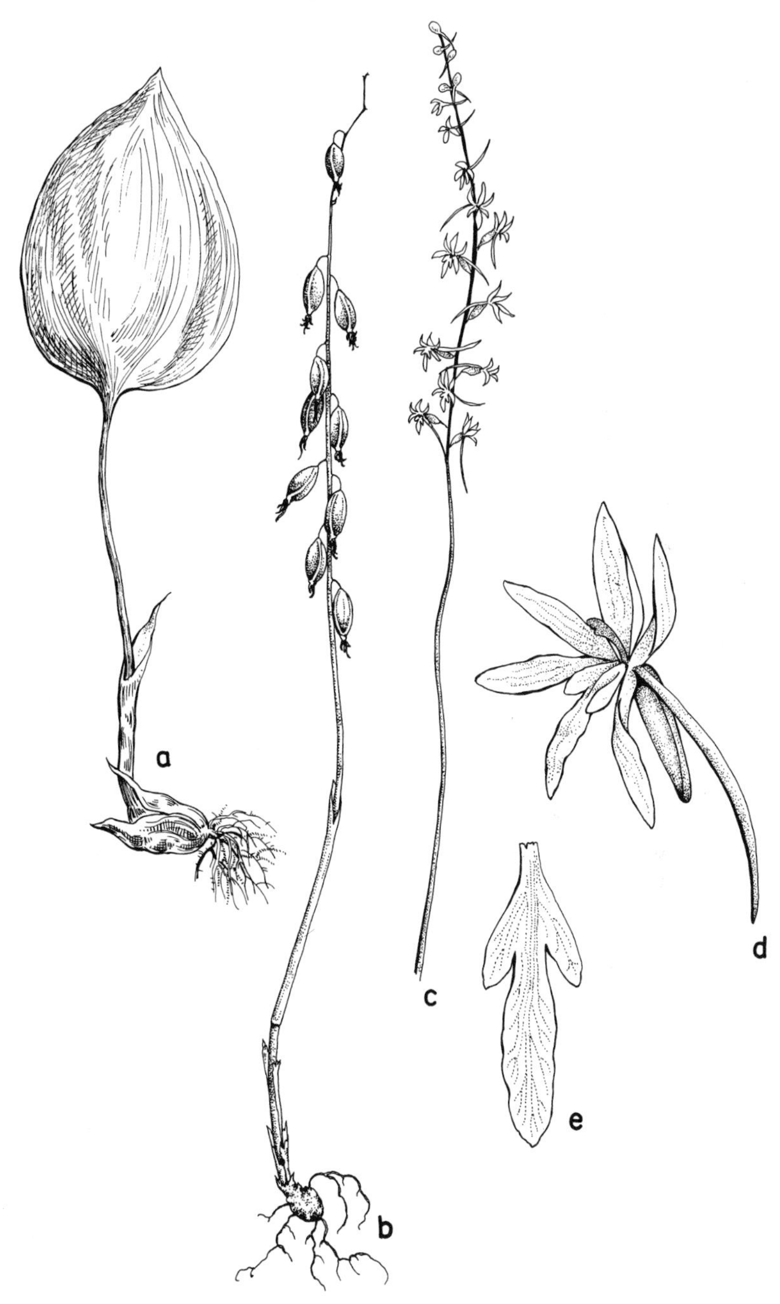

133. *Tipularia discolor* (Crane-fly Orchid). *a.* Leaf, X¼. *b.* Habit, in fruit, X¼. *c.* Inflorescence, X¼. *d.* Flower, X2. *e.* Lip, X4.

COMMON NAME: Crane-fly Orchid.
HABITAT: Rich woodlands.
RANGE: Massachusetts to Illinois, south to Texas and Florida.
ILLINOIS DISTRIBUTION: Very rare; known from extreme southern Illinois. First found in Illinois in 1958 by J. W. Voigt and R. H. Mohlenbrock.

This species may be found most easily in the winter because of the purple-backed, overwintering leaf. The flowers, which bloom in July and August, appear after the leaf has withered. The long spur is another distinguishing character for this species.

Species Excluded

Arethusa bulbosa L. Higley and Raddin (1891) cite a specimen of the swamp-pink orchid collected by Bastin near South Chicago. I have been unable to find this specimen and have therefore omitted this species from the Illinois flora, even though it occurs in Wisconsin and northern Indiana.

Convallaria multiflora L. The report of this species by Mead (1846) is an error for *Polygonatum commutatum* (Schultes) A. Dietr. ex Otto and Dietr.

Cypripedium arietinum R. Br. in Ait. The ram's-head lady's-slipper is recorded from Illinois by Correll (1951) on the basis of a Mead collection, but I have not been able to locate such a specimen. This species generally has a much more northern distribution.

Cypripedium hirsutum Mill. The original report of this species from Illinois by Gates in 1912 is an error for *C. calceolus* var. *parviflorum.*

Goodyera repens R. Br. in Ait. Patterson's report in 1876 of this northern species from Illinois is an error for *G. pubescens* (Willd.) R. Br. in Ait.

Lilium canadense L. This species, which ranges east of Illinois, has been confused with *L. michiganense* Farw. in Illinois.

Lilium catesbaei Walt. Mead first mistakenly credited this southeastern species to Illinois in 1846, and that mistake has erroneously been copied since that time by many authors. This species apparently does not occur in Illinois.

Pancratium rotatum Ker. The genus *Pancratium* sometimes is united with *Hymenocallis.* The binomial *P. rotatum* has been used erroneously by several early workers in Illinois for *H. occidentalis* (LeConte) Kunth.

Smilacina bifolia (L.) Desf. The reference to this species by early Illinois workers was based on misidentifications for *Maianthemum canadense* Desf.

Smilacina trifolia (L.) Desf. Although this species has been attributed to Illinois from Patterson (1876) to Fernald (1950), there is no evidence that it occurs in Illinois. There is a possibility, however, that this species does exist somewhere in northern Illinois.

Spiranthes romanzoffiana Cham. There is no apparent basis for Ames's (1905) or Fuller's (1933) report of this northern ladies-tress from Illinois.

Uvularia lanceolata Ait. Reference to this species from Illinois by Beck in 1826 is an error for *U. grandiflora* Sm.

Veratrum viride Ait. This northern and eastern species was erroneously attributed to Illinois by Snare and Hicks in 1898.

Summary of the Taxa Treated in This Volume

Orders and Families	*Genera*	*Species*	*Varieties*
Order 1. **Liliales**			
Family 1. Liliaceae	30	62	1
Family 2. Smilacaceae	1	9	2
Family 3. Dioscoreaceae	1	2	
Family 4. Iridaceae	4	15	1
Family 5. Burmanniaceae	1	1	
Order 2. **Orchidales**			
Family 6. Orchidaceae	16	43	2
Totals	53	132	6

GLOSSARY

LITERATURE CITED

INDEX OF PLANT NAMES

GLOSSARY

Acaulescent. Seemingly without aerial stems.

Achene. A type of one-seeded, dry, indehiscent fruit with the seed coat not attached to the mature ovary wall.

Acicular. Needle-like.

Actinomorphic. Having radial symmetry; regular, in reference to a flower.

Acuminate. Gradually tapering to a point.

Adnate. Fusion of dissimilar parts.

Alternate. Referring to the condition of structures arising singly along an axis; opposed to opposite.

Annulus. A ring-like structure in the flower of *Thismia.*

Anther. The terminal part of a stamen which bears pollen.

Anthesis. Flowering time.

Antrorse. Projecting forward.

Apically. At the apex.

Apiculate. Abruptly short-pointed at the tip.

Apocarpy. A condition in the flower where more than one free pistil occurs.

Appressed. Lying flat against the surface.

Arborescent. Becoming tree-like.

Arching. Moderately curving.

Areola (pl., **areolae**). A small area between leaf veins.

Aristate. Bearing an awn.

Aristulate. Short-awned.

Attenuate. Gradually becoming narrowed.

Auriculate. Bearing an ear-like process.

Axil. The angle between the base of a structure and the axis from which it arises.

Axile. On the axis, referring to the place of attachment of the ovules.

Axillary. Borne from an axil.

Basal. Confined to the lowest part.

Beak. A terminal projection.

Berry. A type of fruit where the seeds are surrounded only by fleshy material.

Bicostate. Having two veins or ribs.

Biglandular. Bearing two glands.

Bilobed. Bearing two lobes.

Biseriate. In two rows or series.

Bisexual. Referring to a flower which contains both stamens and pistils.

Blade. The green, flat, expanded part of the leaf.

Bract. An accessory structure at the base of many flowers, usually appearing leaflike.

Bracteole. A secondary bract.

Bristle. A stiff hair or hairlike growth; a seta.

Bulb. An underground, vertical stem with both scaly and fleshy leaves.

Bulblet. A small bulb.

Bulbous. Bearing a swollen base.

Callosity. Any hardened thickening.

Calyx. The outermost ring of structures of a flower, composed of sepals.

Campanulate. Bell-shaped.

Capillary. Thread-like.

Capitate. Forming a head.

Capsule. A dry, dehiscent fruit composed of more than one carpel.

Carpel. A simple pistil, or one member of a compound pistil.

Cartilaginous. Firm but flexible.

Caruncle. A fleshy outgrowth near the point of attachment of a seed.

Carunculate. Bearing a fleshy outgrowth near its point of attachment.

Caudate. With a tail-like appendage.

Caudex (pl., **caudices**). The woody base of a perennial plant.

Caulescent. Having an aerial stem.

Cauline. Belonging to a stem.

Cavernous. Hollowed out.

Cespitose. Growing in tufts.

Chartaceous. Papery.

Cilia. Marginal hairs.

Ciliate. Bearing cilia.

Ciliolate. Bearing small cilia.

Clasping. Referring to a leaf whose base encircles the stem.

Claw. A narrow, basal stalk, particularly of a petal.

Compressed. Flattened.

Concave. Curved on the inner surface; opposed to convex.

Connate. Union of like parts.

Connective. That portion of the stamen between the two anther halves.

Connivent. Coming in contact; converging.

Contiguous. Adjoining.

Convex. Rounded on the outer surface; opposite of concave.

Convolute. Rolled lengthwise.

Coralline. Having the texture of coral.

Coralloid. Resembling coral.

Cordate. Heart-shaped.

Coriaceous. Leathery.

Corm. An underground, vertical stem with scaly leaves, differing from a bulb by lacking fleshy leaves.

Corolla. The ring of structures of a flower just within the calyx, composed of petals.

Corona. A crown of petal-like structures, as in *Narcissus*.

Corymb. A type of inflorescence where the pedicellate flowers are arrayed along an elongated axis but with the flowers all attaining about the same height.

Creeping. Spreading on the surface of the ground.

Crested. Bearing a ridge.

Crisped. Curled.

Cross-striae. Markings perpendicular to the longitudinal axis.

Cruciform. Cross-shaped.

Cucullate. Hood-shaped.

Cuneate. Wedge-shaped or tapering at the base.

Cupular. Shaped like a small cup.

Cuspidate. Terminating in a very short point.

Cyme. A type of broad and flattened inflorescence in which the central flowers bloom first.

Cymose. Bearing a cyme.

Deciduous. Falling away.

Decumbent. Lying flat, but with the tip ascending.

Deflexed. Turned downward.

Dehiscent. Splitting at maturity.

Deltoid. Triangular.

Dentate. With sharp teeth, the tips of which project outward.

Denticulate. With small, sharp teeth, the tips of which project outward.

Dilated. Swollen; expanded.

Dimorphic. Having two forms.

Dioecious. With staminate flowers on one plant, pistillate flowers on another.

Distal. Remote from the point of attachment.

Distended. Swollen.

Divaricate. Spreading.

Dorsal. That surface turned away from the axis; abaxial.

Drupaceous. Drupe-like.

Drupe. A type of fruit in which the seed is surrounded by a hard, dry covering which, in turn, is surrounded by fleshy material.

Eglandular. Without glands.

Ellipsoid. Referring to a solid object which is broadest at the middle, gradually tapering to both ends.

Elliptic. Broadest at middle, gradually tapering equally to both ends.

Emersed. Rising above the surface of the water.

Emucronulate. Without a short, abrupt tip.

Endosperm. Food-storage tissue found outside the embryo.

Ensiform. Sword-shaped.

Ephemeral. Lasting only a short time.

Epicarp. The outermost layer of the fruit.

Epunctate. Without dots.

Erose. With an irregularly notched margin.

Exudate. Secreted material.

Facial. Referring to the front surface.

Falcate. Sickle-shaped.

Fascicle. A cluster; a bundle.

Fibrous. Referring to roots borne in tufts.

Filament. That part of the stamen supporting the anther.

Filiform. Threadlike.

Flabellate. Fan-shaped.

Flexuous. Zigzag.

Foliaceous. Leaf-like.

Follicle. A type of dry, dehiscent fruit which splits along one side at maturity.

Friable. Breaking easily into small particles.

Frond. In this volume, the vegetative structure in the Lemnaceae.

Funnelform. Shaped like a funnel.

Fusiform. Spindle-shaped; tapering at both ends.

Galea. A hooded portion of a perianth.

Glabrous. Without pubescence or hairs.

Glaucescent. Becoming covered with a whitish bloom which can be rubbed off.

Glaucous. With a whitish covering which can be rubbed off.

Globoid. Referring to a solid body which is round.

Globose. Round; globular.

Glomerulate. Forming small heads.

Glumaceous. Resembling a scale.

Glutinous. Covered with a sticky secretion.

Hastate. Spear-shaped; said of a leaf which is triangular with spreading basal lobes.

Head. A type of inflorescence in which several sessile flowers are clustered together at the tip of a peduncle.

Hemispherical. Half-spherical.

Herbaceous. Not woody; dying back all the way to the ground in winter.

Hood. That part of an orchid flower which is strongly concave and arching.

Hyaline. Transparent.

Indehiscent. Not splitting open at maturity.

Inferior. Referring to the position of the ovary when it is surrounded by the adnate portion of the floral tube or is embedded in the receptacle.

Inflorescence. A cluster of flowers.

Internode. The area between two adjacent nodes.

Involucral. Referring to a circle of bracts which subtend a flower cluster.

Involute. Rolled inward.

Irregular. In reference to a flower having no symmetry at all.

Keeled. Possessing a ridge-like process.

Lanceolate. Lance-shaped; broadest near base, gradually tapering to the narrow apex.

Lanceoloid. Referring to a solid object which is broadest near base, gradually tapering to the narrow apex.

Linear. Narrow and approximately the same width at either end and the middle.

Lip. The lowermost, often greatly modified petal, in the flower of an orchid.

Lobe. A projection separated from each adjacent projection by a sinus.

Locular. Referring to the cells of a compound ovary.

Locule. A cell or cavity of a compound ovary.

Loculicidal. Said of a capsule which splits down the dorsal suture of each cell.

Lodicule. A membranous scale found within some grass flowers, possibly representing the perianth.

Lustrous. Shiny.

Marginate. With a definite margin.

Mealy. Having a granular texture.

Median. Pertaining to the middle.

Membranous. Like a membrane; thin.

Moniliform. Constricted at regular intervals to resemble a string of beads.

Monoecious. Bearing both sexes in separate flowers on the same plant.

Mucro. A short abrupt tip.

Mucronate. Said of a leaf with a short, terminal point.

Mucronulate. Said of a leaf with a very short, terminal point.

Net-veined. Having veins forming closed meshes.

Node. That place on the stem from which leaves and branchlets arise.

Nodose. Knotty.

Nutlet. A small nut.

Oblanceolate. Reverse lance-shaped; broadest at apex, gradually tapering to narrow base.

Oblique. One-sided; asymmetrical.

Oblong. Broadest at the middle, and tapering to both ends, but broader than elliptic.

Oblongoid. Referring to a solid object which, in side view, is nearly the same width throughout, but broader than linear.

Obovoid. Referring to a solid object which is broadly rounded at the apex, becoming narrowed below.

Obpyramidal. Referring to an upside-down pyramid.

Obtuse. Rounded at the apex.

Opaque. Incapable of being seen through.

Opposite. Referring to the condition of two like structures arising from the same point.

Orbicular. Round.

Orthotropous. Referring to an ovule which is borne upright.

Ovary. The lower swollen part of the pistil which produces the ovules.

Ovate. Broadly rounded at base, becoming narrowed above; broader than lanceolate.

Ovoid. Referring to a solid object which is broadly rounded at the base, becoming narrowed above.

Ovule. The egg-producing structure found within the ovary.

Panduriform. Fiddle-shaped.

Panicle. A type of inflorescence composed of several racemes.

Papillose. Bearing pimple-like processes.

Papule. A pimple-like projection.

Parallel-veined. Having veins running in the same direction and not meeting.

Pedicel. The stalk of a flower of an inflorescence.

Peduncle. The stalk of an inflorescence.

Pedunculate. Bearing a peduncle.

Pellucid. Being transparent, in reference to spots or dots.

Peltate. Attached away from the margin, in reference to a leaf.

Pendent. Suspended; overhanging.

Pendulous. Hanging.

Perennial. Living more than two years.

Perfect. Bearing both stamens and pistils in the same flower.

Perfoliate. Referring to a leaf which appears to have the stem pass through it.

Perianth. Those parts of a flower including both the calyx and corolla.

Persistent. Remaining attached.

Petal. One segment of the corolla.

Petaloid. Resembling a petal in texture and appearance.

Petiolate. Bearing a petiole, or leafstalk.

Petiole. The stalk of a leaf.

Phyllodia. Dilated petioles modified to resemble and function as leaves.

Pilose. Bearing soft hairs.

Pilosulous. Bearing short, soft hairs.

Pistil. The ovule-producing organ of a flower normally composed of an ovary, a style, and a stigma.

Pistillate. Bearing pistils but not stamens.

Placentation. Referring to the manner in which the ovules are attached in the ovary.

Plicate. Folded.

Podogyne. A stalk in *Ruppia* on top of which the fruit is produced.

Prophyll. A bracteole.

Prophyllate. Bearing a bracteole at the base of a flower.

Prostrate. Lying flat.

Puberulent. With minute hairs.

Pubescent. Bearing some kind of hairs.

Punctation. A dot or dots.

Quadrangular. Four-angled.

Quadrate. Four-sided.

Raceme. A type of inflorescence where pedicellate flowers are arranged along an elongated axis.

Ranked. Referring to the number of planes in which structures are borne.

Receptacle. That part of the flower to which the perianth, stamens, and pistils are usually attached.

Reflexed. Turned downward.

Regular. Having radial symmetry (actinomorphic) or bilateral symmetry (zygomorphic).

Reniform. Kidney-shaped.

Resin. A usually sticky secretion found in various parts of certain plants.

Resupinate. Upside down.

Reticulate. Resembling a network.

Retuse. Shallowly notched at a rounded apex.

Rhizome. An underground horizontal stem, bearing nodes, buds, and roots.

Rhombic. Becoming quadrangular.

Ribbed. Nerved; veined.

Root cap. A group of cells borne externally at the tip of a root.

Rosette. A cluster of leaves in a circular arrangement at the base of a plant.

Rotate. Flat and circular.

Rugose. Wrinkled.

Rugulose. With small wrinkles.

Saccate. Sac-shaped.

Sagittate. Shaped like an arrowhead.

Salverform. Referring to a tubular corolla which abruptly expands into a flat limb.

Saprophyte. A type of plant living on dead or decaying organic matter, usually through the medium of mycorrhizae.

Scaberulous. Minutely roughened; slightly rough to the touch.

Scabrous. Rough to the touch.

Scale. A minute epidermal outgrowth, sometimes becoming green and replacing the leaf in function.

Scape. A leafless stalk bearing a flower or inflorescence.

Scapose. Possessing a leafless flowering stem.

Scarious. Thin and membranous.

Scurfy. Bearing scaly particles.

Secund. Borne on one side.

Sepal. One segment of the calyx.

Sepaloid. Appearing like a sepal.

Septate. With cross walls.

Septicidal. Said of a capsule which splits between the locules.

Serrulate. With very small teeth, the tips of which project forward.

Sessile. Without a stalk.

Setaceous. Bearing bristles, or setae.

Setose. Bearing bristles.

Sheath. A protective covering.

Shoot. The developing stem with its leaves.

Spadix. A fleshy axis in which flowers are embedded.

Spathe. A large sheathing bract subtending or usually enclosing an inflorescence.

Spatulate. Oblong, but with the basal end elongated.

Spike. A type of inflorescence where sessile flowers are arranged along an elongated axis.

Spinescent. Becoming spiny.

Spinule. A small spine.

Spinulose. With small spines.

Stamen. The pollen-producing organ of a flower composed of a filament and an anther.

Staminate. Bearing stamens but not pistils.

Staminodium (pl., **staminodia**). A sterile stamen.

Stipitate. Bearing a stipe or stalk.

Stipule. A leaf-like or scaly structure found at the point of attachment of a leaf to the stem.

Stoloniferous. Bearing runners or slender horizontal stems on the surface of the ground.

Stramineous. Straw-colored.

Style. That part of the pistil between the ovary and the stigma.

Subacute. Nearly acute.

Subcapitate. Nearly head-like.

Subligneous. Nearly woody.

Submersed. Covered with water.

Suborbiculate. Nearly round.

Subsessile. Nearly without a stalk.

Subulate. Drawn to an abrupt sharp point.

Subuloid. Referring to a solid object which is drawn to an abrupt short point.

Suffused. Spread throughout; flushed.

Superior. Referring to the position of the ovary when the free floral parts arise below the ovary.

Supra-axillary. Borne above the axil.

Tendril. A spiralling, coiling structure which enables a climbing plant to attach itself to a supporting body.

Tenuous. Slender.

Terete. Rounded in cross-section.

Thalloid. Possessing an undifferentiated plant body, that is, without roots or stems or leaves.

Throat. The opening at the apex of a corolla tube where the limb arises.

Translucent. Partly transparent.

Trilocular. With three cavities.

Trimorphic. Having three forms.

Truncate. Abruptly cut across.

Tuber. An underground fleshy stem formed as a storage organ at the end of a rhizome.

Tubercle. A small wart-like process.

Tubular. Shaped like a tube.

Tunicated. Covered with concentric coats.

Turbinate. Shaped like a top.

Turgid. Tightly inflated.

Umbel. A type of inflorescence in which the flower stalks arise from the same level.

Umbonate. With a stout projection at the center.

Unarmed. Without prickles or spines.

Undulate. Wavy.

Unisexual. Bearing either stamens or pistils in one flower.

Utricle. A small, one-seeded, indehiscent fruit with a thin covering.

Valve. That part of a capsule which splits.

Venation. The pattern of the veins.

Ventral. That surface turned toward the axis; adaxial.

Villous. With long soft erect hairs.

Viscid. Sticky.

Whorl. An arrangement of three or more structures at a point on the stem.

Whorled. Referring to the condition of three or more like structures arising from the same point.

Winged. Bearing a flat lateral outgrowth.

Zygomorphic. Bilaterally symmetrical.

LITERATURE CITED

Akemine, T. 1935. Chromosome studies on *Hosta* I. The chromosome number in various species of *Hosta*. Journal of the Faculty of Science of Hokkaido University 5:25–32.

Ames, O. 1905. A synopsis of the genus *Spiranthes* north of Mexico, in Orchidaceae 1:122–54.

———. 1924. An enumeration of the orchids of the United States and Canada. Boston, Massachusetts. 120 pp.

Anderson, E. & T. W. Whitaker. 1934. Speciation in *Uvularia*. Journal of the Arnold Arboretum 15:28–42.

Bamford, R. 1935. The chromosome number in *Gladiolus*. Journal of Agricultural Research 51:945–50.

Beck, L. C. 1826. Contributions towards the botany of the states of Illinois and Missouri. American Journal of Science and Arts 10:257–64; 11:167–82.

Bicknell, E. P. 1896. The blue-eyed grasses of the eastern United States (Genus *Sisyrinchium*). Bulletin of the Torrey Botanical Club 23:130–37.

Carlson, M. C. 1945. Megasporogenesis and development of the embryo sac of *Cypripedium parviflorum*. Botanical Gazette 107: 107–14.

Case, F. W. 1964. Orchids of the Western Great Lakes Region. Cranbrook Institute of Science Bulletin 48. Bloomfield Hills, Michigan. 147 pp.

Cooper, D. C. 1939. Development of megagametophyte in *Erythronium albidum*. Botanical Gazette 100:862–67.

Correll, D. S. 1950. Native orchids of North America. Chronica Botanica Company, Waltham, Massachusetts. 399 pp.

Cronquist, A. 1968. The evolution and classification of flowering plants. Houghton Mifflin Company, Boston. 396 pp.

Dark, S. O. S. 1932. Meiosis in diploid and triploid *Hemerocallis*. New Phytologist 31 (5):310–20.

Darlington, C. D. and L. F. LaCour. 1940. Nucleic acid starvation of chromosomes in *Trillium*. Journal of Genetics 40:185–213.

Deam, C. C. 1940. Flora of Indiana. Indianapolis, Indiana. 1236 pp.

Delaunay, L. N. 1926. Chromosome theory of heredity and chromosomes in certain Liliaceae. Moniteur Jard. Bot. Tiflis 2:1–32.

Farwell, O. A. 1928. Botanical gleanings in Michigan. V. The American Midland Naturalist 11:41–71.

Fernald, M. L. 1946. *Stenanthium* in the Eastern United States. Rhodora 48:148–52.

Fernald, M. L. 1950. Gray's Manual of Botany. Edition 8. The American Book Company, New York. 1632 pp.

Fuller, A. M. 1933. Studies on the flora of Wisconsin. Part I. The Orchids; Orchidaceae. Bulletin of the Public Museum of Milwaukee 14:1–284.

Gadella, T. W. J. and K. Kliphuis. 1963. Chromosome numbers of flowering plants in the Netherlands. Acta Bot. Neerl. 12:195–230.

Gates, F. C. 1912. The vegetation of the beach area in northeastern Illinois and southeastern Wisconsin. Bulletin of the Illinois State Laboratory of Natural History 9:255–372.

Hanes, C. R. 1953. *Allium tricoccum* Ait., var. *Burdickii,* var. nov. Rhodora 55:243–44.

———, and M. Ownbey. 1946. Some observations on two ecological races of *Allium tricoccum* in Kalamazoo County, Michigan. Rhodora 48:61–63.

Haque, A. 1951. Embryo sac of *Erythronium americanum.* Botanical Gazette 112:495–500.

Higley, W. K. and C. S. Raddin. 1891. Flora of Cook County, Illinois and a part of Lake County, Indiana. Bulletin of the Chicago Academy of Science 2:1–168.

Hitchcock, C. L. 1944. The *Tofieldia glutinosa* complex of western North America. The American Midland Naturalist 31:487–98.

Humphrey, L. M. 1932. Somatic chromosomes of certain Minnesota orchids. The American Naturalist 66:471–74.

Ingram, J. 1968. Notes on the cultivated Liliaceae 7. *Lilium lancifolium* Thunb. vs. *L. tigrinum* Ker. Baileya 16:14–19.

Kibbe, A. 1952. A botanical study and survey of a typical midwestern county (Hancock County, Illinois). Carthage, Illinois. 425 pp.

Levan, A. 1931. Cytological studies in *Allium.* A preliminary note. Hereditas 15:347–56.

———. 1935. Cytological studies in *Allium,* VI. The chromosome morphology of some diploid species of *Allium.* Hereditas 20:287–330.

———. 1936. Zytologische studien an *Allium schoenoprasum.* Hereditas 22:1–128.

Löve, A. and D. Löve. 1948. Chromosome numbers of northern plant species. Ingolfsprent, Reykjavik, Iceland.

Mangaly, J. 1968. A cytotaxonomic study of the herbaceous species of *Smilax:* Section Coprosmanthus. Rhodora 70:55–82, 247–73.

Matsuura, H. and T. Sutô. 1935. Contributions to the idiogram study in phanerogamous plants. Journal of the Faculty of Science of Hokkaido University 5:33–75.

McKelvey, S. D. and K. Sax. 1933. Taxonomic and cytological relationships of *Yucca* and *Agave.* Journal of the Arnold Arboretum 14:76–81.

Mead, S. B. 1846. Catalogue of plants growing spontaneously in the State of Illinois, the principal part near Augusta, Hancock County. Prairie Farmer 6:35–36, 60, 93, 119–22.

Mohlenbrock, R. H. 1954 Flowering plants and ferns of Giant City State Park. Division of Parks and Memorials and the Illinois State Museum, Springfield. 24 pp.

———. 1962. On the occurrence of *Lilium superbum* in Illinois. Castanea 27:173–76.

Ownbey, M. and H. C. Aase. 1955. Cytotaxonomic studies in *Allium*. I. The *Allium canadense* alliance. Research Studies of the State College of Washington 4 (Suppl.):1–106.

Ownbey, R. P. 1944. The liliaceous genus *Polygonatum* in North America. Annals of the Missouri Botanical Garden 31:373–413.

Parks, C. R. and J. W. Hardin. 1963. Yellow Erythroniums of the Eastern United States. Brittonia 15:245–59.

Patterson, H. N. 1876. Catalogue of the phaenogamous and vascular cryptogamous plants of Illinois. Oquawka, Illinois. 54 pp.

Pepoon, H. S. 1909. The cliff flora of JoDaviess County. Transactions of the Illinois State Academy of Science 2:32–37.

Pepoon, H. S. 1927. An annotated flora of the Chicago area. Bulletin of the Chicago Academy of Science 8:1–554.

Sansome, E. R. and L. F. LaCour. 1934. Lily Yearbook 40.

Sato, M. 1932. Chromosome studies in *Lilium*. I. Botanical Magazine of Tokyo 46:68–88.

Shinners, L. H. 1966. Texas *Polianthes*, including *Manfreda* (*Agave* Subgenus Manfreda) and *Runyonia* (Agavaceae). Sida 2 (4):333–38.

Smith, B. W. 1937. Notes on the cytology and distribution of the Dioscoreaceae. Bulletin of the Torrey Botanical Club 64:189–97.

Snare, W. and E. W. Hicks. 1898. Check list of plants in the Boardman Collection, Toulon Academy. Privately published. 29 pp.

Sokolovskaya, A. P. 1963. Geographical distribution of polyploidy in plants. Vest. Leningrad Univ. No. 15, Ser. Biol.:38–52.

Sorsa, V. 1962. Chromosomenzahlen Finnischer Kormophyton I. Ann. Acad. Sci. Fenn. Ser. A. IV. Biol. 58:1–14.

———. 1963. Chromosomenzahlen Finnischer Kormophyton II. Ann. Acad. Sci. Fenn. Ser. A. IV. Biol. 68:1–14.

Stewart, R. N. 1947. The morphology of somatic chromosomes in *Lilium*. American Journal of Botany 34:9–26.

———, and R. Bamford. 1942. The chromosomes and nucleoli of *Medeola virginiana*. American Journal of Botany 29:301–3.

Steyermark, J. A. 1961. A neglected *Camassia*. Brittonia 13:212–24.

———, and F. W. Swink. 1955. Plants new to Illinois and to the Chicago region. Rhodora 57:265–68.

Stout, A. B. 1932. Chromosome numbers in *Hemerocallis* with reference to triploidy and secondary polyploidy. Cytologia 3:250–59.

Thorne, R. F. 1968. Synopsis of a putatively phylogenetic classification of the flowering plants. Aliso 6 (4):57–66.

Ward, D. B. 1968. The nomenclature of *Sisyrinchium bermudiana* and related North American species. Taxon 17:270–76.

Wherry, E. T. 1942. The relationship of *Lilium michiganense*. Rhodora 44:453–56.

Wilbur, R. 1963. A revision of the North American genus *Uvularia* (Liliaceae). Rhodora 65:158–88.

Winterringer, G. S. 1967. Wild orchids of Illinois. Illinois State Museum, Springfield. 130 pp.

Wunderlich, R. 1937. Zur vergleichenden embryologie der Liliaceae-Scillioideae. Flora 32:48–90.

INDEX OF PLANT NAMES